Introductory Geography

Introductory Geography

Christopher Zimmer

SYRAWOOD
PUBLISHING HOUSE

New York

Published by Syrawood Publishing House,
750 Third Avenue, 9th Floor,
New York, NY 10017, USA
www.syrawoodpublishinghouse.com

Introductory Geography
Christopher Zimmer

International Standard Book Number: 978-1-64740-018-7 (Hardback)

Cataloging-in-Publication Data

Introductory geography / Christopher Zimmer.
    p. cm.
Includes bibliographical references and index.
ISBN 978-1-64740-018-7
1. Geography. 2. Earth sciences. 3. Cosmography. I. Zimmer, Christopher.
G116 .I58 2020
910--dc23

TABLE OF CONTENTS

Permissions

Index

PREFACE

The field of study, which deals with features, lands, inhabitants and phenomena of the Earth and planets is known as geography. It seeks to understand the Earth, along with the natural and human complexities. The two broad divisions of geography are human geography and physical geography. The people and their communities, as well as their cultures and interactions with the environment are studied within human geography. The different processes and patterns in the natural environment such as the hydrosphere, biosphere, geosphere and atmosphere are studied within physical geography. Some of the tools used within this field of study are cartography, geographic information systems and remote sensing. This book provides significant information of this discipline to help develop a good understanding of geography and related fields. It includes contributions of experts and scientists which will provide innovative insights into this field. The book aims to serve as a resource guide for students and experts alike and contribute to the growth of the discipline.

A foreword of all Chapters of the book is provided below:

Chapter 1 - The field of science which deals with the study of lands, inhabitants, features and the phenomena concerning Earth and planets is known as geography. Some of the different branches of geography are physical geography, human geography and integrated geography. This is an introductory chapter which will introduce briefly all the significant aspects of geography.; **Chapter 2** - The formation of Earth happened roughly 4.54 billion years ago, which is roughly two thirds of the age of the universe. The two models which seek to explain the formation of the Earth are the core accretion model and the disk instability model. The chapter closely examines these key concepts related to the formation of the Earth to provide an extensive understanding of the subject.; **Chapter 3** - A solid compound which occurs naturally is known as a mineral. Rocks are solid masses which are comprised of minerals. Some of the properties of minerals, on the basis of which they can be classified are color, hardness and tenacity. All these diverse properties of minerals and rocks have been carefully analyzed in this chapter.; **Chapter 4** - Geomorphic processes are the endogenic and exogenic forces which cause chemical actions and physical stresses on earth materials. Some of the geomorphic processes are mass movements, erosion and deposition, and soil formation. The topics elaborated in this chapter will help in gaining a better perspective about these types of geomorphic processes.; **Chapter 5** - The small to medium tracts of Earth's surface are called landforms. The evolution of landforms occurs due to interaction between different physical processes and environmental factors, such as tectonics, underlying rock structures, climate and climatic changes, rock types, and human activities. This chapter discusses in detail these theories and methodologies related to landforms and their evolution.; **Chapter 6** - The total number of people which reside in a particular place over a period of time constitute the population of that place. The study of populations involves studying its composition, density and growth. The chapter closely examines these key concepts of population and migration to provide an extensive understanding of the subject.; **Chapter 7** - The process which is involved in enlarging the freedom of people and improving their well-being is known as human development. One of the ways in which it is measured is through human development index. These diverse aspects of human development in the current scenario have been thoroughly discussed in this chapter.; **Chapter 8** - The branch of science which is involved in the study of all the aspects of ocean is known as oceanography. Some of the areas of study within this discipline are ocean temperature, ocean salinity and ocean floor relief. The chapter closely examines these key aspects of oceanography to provide an extensive understanding of the subject.

I would like to thank the entire editorial team who made sincere efforts for this book and my family who supported me in my efforts of working on this book. I take this opportunity to thank all those who have been a guiding force throughout my life.

Christopher Zimmer

Chapter 1

Introduction to Geography

The field of science which deals with the study of lands, inhabitants, features and the phenomena concerning Earth and planets is known as geography. Some of the different branches of geography are physical geography, human geography and integrated geography. This is an introductory chapter which will introduce briefly all the significant aspects of geography.

Geography is the study of places and the relationships between people and their environments. Geographers explore both the physical properties of Earth's surface and the human societies spread across it. They also examine how human culture interacts with the natural environment and the way that locations and places can have an impact on people. Geography seeks to understand where things are found, why they are there, and how they develop and change over time.

The term "geography" comes to us from the ancient Greeks, who needed a word to describe the writings and maps that were helping them make sense of the world in which they lived. In Greek, geo means "earth" and -graphy means "to write." Using geography, Greeks developed an understanding of where their homeland was located in relation to other places, what their own and other places were like, and how people and environments were distributed. These concerns have been central to geography ever since.

Earth

Of course, the Greeks were not the only people interested in geography. Throughout human history, most societies have sought to understand something about their place in the world, and the people and environments around them.

Indeed, mapmaking probably came even before writing in many places. But ancient Greek geographers were particularly influential. They developed very detailed maps of areas in and around Greece, including parts of Europe, Africa, and Asia. More importantly, they also raised questions about how and why different human and natural patterns came into being on Earth's surface, and why variations existed from place to place. The effort to answer these questions about patterns and distribution led them to figure out that the world was round, to calculate Earth's circumference, and to develop explanations of everything from the seasonal flooding of the Nile River to differences in population densities from place to place.

During the Middle Ages, geography ceased to be a major academic pursuit in Europe. Advances in geography were chiefly made by scientists of the Muslim world, based around the Arabian Peninsula and North Africa. Geographers of this Islamic Golden Age created the world's first rectangular map based on a grid, a map system that is still familiar today. Islamic scholars also applied their study of people and places to agriculture, determining which crops and livestock were most suited to specific habitats or environments.

Marco polo's caravan.

In addition to the advances in the Middle East, the Chinese empire in Asia also contributed immensely to geography. Until about 1500, China was the most prosperous civilization on Earth. The Chinese were scientifically advanced, especially in the field of astronomy. Around 1000, they also achieved one of the most important developments in the history of geography: They were the first to use the compass for navigational purposes. In the early 1400s, the explorer Cheng Ho embarked on seven voyages to the lands bordering the China Sea and the Indian Ocean, establishing China's dominance throughout Southeast Asia.

Age of Discovery

Through the 13th-century travels of the Italian explorer Marco Polo, Europeans learned about the riches of China. Curiosity was awakened; a desire to trade with wealthy Asian cultures motivated a renewed interest in exploring the world. The period of time between the 15th and 17th centuries is known in the West as the Age of Exploration or the Age of Discovery.

With the dawn of the Age of Discovery, the study of geography regained popularity in Europe. The invention of the printing press in the mid-1400s helped spread geographic knowledge by making maps and charts widely available. Improvements in shipbuilding and navigation facilitated more exploring, greatly improving the accuracy of maps and geographic information.

Greater geographic understanding allowed European powers to extend their global influence. During the Age of Discovery, European nations established colonies around the world. Improved transportation, communication, and navigational technology allowed countries such as the United Kingdom to successfully govern colonies as far away as the Americas, Asia, Australia, and Africa.

Geography was not just a subject that made colonialism possible, however. It also helped people understand the planet on which they lived. Not surprisingly, geography became an important focus of study in schools and universities.

Geography also became an important part of other academic disciplines, such as chemistry, economics, and philosophy. In fact, every academic subject has some geographic connection. Chemists study where certain chemical elements, such as gold or silver, can be found. Economists examine which nations trade with other nations, and what resources are exchanged. Philosophers analyze the responsibility people have to take care of the Earth.

Emergence of Modern Geography

Some people have trouble understanding the complete scope of the discipline of geography because, unlike most other disciplines, geography is not defined by one particular topic. Instead, geography is concerned with many different topics—people, culture, politics, settlements, plants, landforms, and much more.

What distinguishes geography is that it approaches the study of diverse topics in a particular way (that is, from a particular perspective). Geography asks spatial questions—how and why things are distributed or arranged in particular ways on Earth's surface. It looks at these different distributions and arrangements at many different scales. It also asks questions about how the interaction of different human and natural activities on Earth's surface shape the characteristics of the world in which we live.

Geography seeks to understand where things are found and why they are present in those places; how things that are located in the same or distant places influence one another over time; and why places and the people who live in them develop and change in particular ways. Raising these questions is at the heart of the "geographic perspective."

Exploration has long been an important part of geography. But exploration no longer simply means going to places that have not been visited before. It means documenting and trying to explain the variations that exist across the surface of Earth, as well as figuring out what those variations mean for the future.

The age-old practice of mapping still plays an important role in this type of exploration, but exploration can also be done by using images from satellites or gathering information from interviews. Discoveries can come by using computers to map and analyze the relationship among things in geographic space, or from piecing together the multiple forces, near and far, that shape the way individual places develop.

Applying a geographic perspective demonstrates geography's concern not just with where things are, but with "the why of where"—a short, but useful definition of geography's central focus.

The insights that have come from geographic research show the importance of asking "the why of where" questions. Geographic studies comparing physical characteristics of continents on either side of the Atlantic Ocean, for instance, gave rise to the idea that Earth's surface is comprised of large, slowly moving plates—plate tectonics.

Studies of the geographic distribution of human settlements have shown how economic forces and modes of transport influence the location of towns and cities. For example, geographic analysis has pointed to the role of the U.S. Interstate Highway System and the rapid growth of car ownership in creating a boom in U.S. suburban growth after World War II. The geographic perspective helped show where Americans were moving, why they were moving there, and how their new living places affected their lives, their relationships with others, and their interactions with the environment.

Geographic analyses of the spread of diseases have pointed to the conditions that allow particular diseases to develop and spread. Dr. John Snow's cholera map stands out as a classic example. When cholera broke out in London, England, in 1854, Snow represented the deaths per household on a street map. Using the map, he was able to trace the source of the outbreak to a water pump on the corner of Broad Street and Cambridge Street. The geographic perspective helped identify the source of the problem (the water from a specific pump) and allowed people to avoid the disease (avoiding water from that pump).

Investigations of the geographic impact of human activities have advanced understanding of the role of humans in transforming the surface of Earth, exposing the spatial extent of threats such as water pollution by manmade waste. For example, geographic study has shown that a large mass of tiny pieces of plastic currently floating in the Pacific Ocean is approximately the size of Texas. Satellite images and other geographic technology identified the so-called "Great Pacific Garbage Patch."

These examples of different uses of the geographic perspective help explain why geographic study and research is important as we confront many 21st century challenges, including environmental pollution, poverty, hunger, and ethnic or political conflict.

Because the study of geography is so broad, the discipline is typically divided into specialties. At the broadest level, geography is divided into physical geography, human geography, geographic techniques, and regional geography.

Physical Geography

The natural environment is the primary concern of physical geographers, although many physical geographers also look at how humans have altered natural systems. Physical geographers study Earth's seasons, climate, atmosphere, soil, streams, landforms, and oceans. Some disciplines within physical geography include geomorphology, glaciology, pedology, hydrology, climatology, biogeography, and oceanography.

Geomorphology is the study of landforms and the processes that shape them. Geomorphologists investigate the nature and impact of wind, ice, rivers, erosion, earthquakes, volcanoes, living things, and other forces that shape and change the surface of the Earth.

Glaciologists focus on the Earth's ice fields and their impact on the planet's climate. Glaciologists document the properties and distribution of glaciers and icebergs. Data collected by glaciologists has demonstrated the retreat of Arctic and Antarctic ice in the past century.

Pedologists study soil and how it is created, changed, and classified. Soil studies are used by a variety of professions, from farmers analyzing field fertility to engineers investigating the suitability of different areas for building heavy structures.

Hydrology is the study of Earth's water: its properties, distribution, and effects. Hydrologists are especially concerned with the movement of water as it cycles from the ocean to the atmosphere, then back to Earth's surface. Hydrologists study the water cycle through rainfall into streams, lakes, the soil, and underground aquifers. Hydrologists provide insights that are critical to building or removing dams, designing irrigation systems, monitoring water quality, tracking drought conditions, and predicting flood risk.

Climatologists study Earth's climate system and its impact on Earth's surface. For example, climatologists make predictions about El Nino, a cyclical weather phenomenon of warm surface temperatures in the Pacific Ocean. They analyze the dramatic worldwide climate changes caused by El Nino, such as flooding in Peru, drought in Australia, and, in the United States, the oddities of heavy Texas rains or an unseasonably warm Minnesota winter.

Biogeographers study the impact of the environment on the distribution of plants and animals. For example, a biogeographer might document all the places in the world inhabited by a certain spider species, and what those places have in common.

Oceanography, a related discipline of physical geography, focuses on the creatures and environments of the world's oceans. Observation of ocean tides and currents constituted some of the first oceanographic investigations. For example, 18th-century mariners figured out the geography of the Gulf Stream, a massive current flowing like a river through the Atlantic Ocean. The discovery and tracking of the Gulf Stream helped communications and travel between Europe and the Americas.

Today, oceanographers conduct research on the impacts of water pollution, track tsunamis, design offshore oil rigs, investigate underwater eruptions of lava, and study all types of marine organisms from toxic algae to friendly dolphins.

Human Geography

Human geography is concerned with the distribution and networks of people and cultures on Earth's surface. A human geographer might investigate the local, regional, and global impact of rising economic powers China and India, which represent 37 percent of the world's people. They also might look at how consumers in China and India adjust to new technology and markets, and how markets respond to such a huge consumer base.

Human geographers also study how people use and alter their environments. When, for example, people allow their animals to overgraze a region, the soil erodes and grassland is transformed into desert. The impact of overgrazing on the landscape as well as agricultural production is an area of study for human geographers.

Finally, human geographers study how political, social, and economic systems are organized across geographical space. These include governments, religious organizations, and trade partnerships. The boundaries of these groups constantly change.

The main divisions within human geography reflect a concern with different types of human activities or ways of living. Some examples of human geography include urban geography, economic geography, cultural geography, political geography, social geography, and population geography. Human geographers who study geographic patterns and processes in past times are part of the subdiscipline of historical geography. Those who study how people understand maps and geographic space belong to a subdiscipline known as behavioral geography.

Many human geographers interested in the relationship between humans and the environment work in the subdisciplines of cultural geography and political geography.

Cultural geographers study how the natural environment influences the development of human culture, such as how the climate affects the agricultural practices of a region. Political geographers study the impact of political circumstances on interactions between people and their environment, as well as environmental conflicts, such as disputes over water rights.

Some human geographers focus on the connection between human health and geography. For example, health geographers create maps that track the location and spread of specific diseases. They analyze the geographic disparities of health-care access. They are very interested in the impact of the environment on human health, especially the effects of environmental hazards such as radiation, lead poisoning, or water pollution.

Sing-Sing

Geographic Techniques

Specialists in geographic techniques study the ways in which geographic processes can be analyzed and represented using different methods and technologies. Mapmaking, or cartography, is perhaps the most basic of these. Cartography has been instrumental to geography throughout the ages.

As early as 1500 BCE, Polynesian navigators in the Pacific Ocean used complex maps made of tiny sticks and shells that represented islands and ocean currents they would encounter on their voyages. Today, satellites placed into orbit by the U.S. Department of Defense communicate with receivers on the ground called global positioning system (GPS) units to instantly identify exact locations on Earth.

Today, almost the entire surface of Earth has been mapped with remarkable accuracy, and much of this information is available instantly on the internet. One of the most remarkable of these websites is Google Earth, which "lets you fly anywhere on Earth to view satellite imagery, maps, terrain, 3D buildings, from galaxies in outer space to the canyons of the ocean." In essence, anyone can be a virtual Christopher Columbus from the comfort of home.

Technological developments during the past 100 years have given rise to a number of other specialties for scientists studying geographic techniques. The airplane made it possible to photograph land from above. Now, there are many satellites and other above-Earth vehicles that help geographers figure out what the surface of the planet looks like and how it is changing.

Geographers looking at what above-Earth cameras and sensors reveal are specialists in remote sensing. Pictures taken from space can be used to make maps, monitor ice melt, assess flood damage, track oil spills, predict weather, or perform endless other functions. For example, by comparing satellite photos taken from 1955 to 2007, scientists from the U.S. Geological Survey (USGS) discovered that the rate of coastal erosion along Alaska's Beaufort Sea had doubled. Every year from 2002 to 2007, about 45 feet per year of coast, mostly icy permafrost, vanished into the sea.

Computerized systems that allow for precise calculations of how things are distributed and relate to one another have made the study of geographic information systems (GIS) an increasingly important specialty within geography. Geographic information systems are powerful databases that collect all types of information (maps, reports, statistics, satellite images, surveys, demographic data, and more) and link each piece of data to a geographic reference point, such as geographic coordinates. This data, called geospatial information, can be stored, analyzed, modeled, and manipulated in ways not possible before GIS computer technology existed.

The popularity and importance of GIS has given rise to a new science known as geographic information science (GISci). Geographic information scientists study patterns in nature as well as human development. They might study natural hazards, such as a fire that struck Los Angeles, California, in 2008. The real-time spread of the fire along with information to help people make decisions about how to evacuate quickly. GIS can also illustrate human struggles from a geographic perspective, such as the interactive map that showed building foreclosure rates in various regions.

The enormous possibilities for producing computerized maps and diagrams that can help us understand environmental and social problems have made geographic visualization an increasingly important specialty within geography. This geospatial information is in high demand by just about every institution, from government agencies monitoring water quality to entrepreneurs deciding where to locate new businesses.

Regional Geography

Regional geographers take a somewhat different approach to specialization, directing their attention to the general geographic characteristics of a region. A regional geographer might specialize in African studies, observing and documenting the people, nations, rivers, mountains, deserts, weather, trade, and other attributes of the continent. There are different ways you can define a region. You can look at climate zones, cultural regions, or political regions. Often regional geographers have a physical or human geography specialty as well as a regional specialty.

Regional geographers may also study smaller regions, such as urban areas. A regional geographer may be interested in the way a city like Shanghai, China, is growing. They would study transportation, migration, housing, and language use, as well as the human impact on elements of the natural environment, such as the Huangpu River.

International Space Station.

Danxia Landform.

Whether geography is thought of as a discipline or as a basic feature of our world, developing an understanding of the subject is important. Some grasp of geography is essential as people seek to make sense of the world and understand their place in it. Thinking geographically helps people to be aware of the connections among and between places and to see how important events are shaped by where they take place. Finally, knowing something about geography enriches people's lives—promoting curiosity about other people and places and an appreciation of the patterns, environments, and peoples that make up the endlessly fascinating, varied planet on which we live.

March of two Penguins.

Scuba Diving.

Chapter 2

Formation of the Earth

The formation of Earth happened roughly 4.54 billion years ago, which is roughly two thirds of the age of the universe. The two models which seek to explain the formation of the Earth are the core accretion model and the disk instability model. The chapter closely examines these key concepts related to the formation of the Earth to provide an extensive understanding of the subject.

Evolution

The Big Bang and Atomic Synthesis

The universe is thought to have begun as a tiny package containing all matter which burst apart about 14 billion years ago in what is known as "The Big Bang". It is still expanding from this initial explosion. What happened before the Big Bang is unknown as is the fate of the universe – whether it will continue to expand, or whether gravitational forces will overcome the expansion and begin to recall the material to the center of mass perhaps to explode again. (Current observational evidence suggests that there is not enough mass to stop expansion though it is still possible that astronomers will find some previously-unknown mass sufficient to cause expansion to stop).

We know the age of the universe and that it is expanding from examination of the light spectra coming to us from distant objects in the universe. The light is shifted to longer wave lengths (the "red shift"). This can be explained as a Doppler shift caused by the fact that the objects are moving away from us.

For some time after the Big Bang, the universe consisted only of gaseous hydrogen and helium – there were no stars or galaxies. All other elements were created during the life and death of stars. Normal stellar evolution produces only elements up to iron and so the heavier elements must have formed inside stars which subsequently exploded ("supernovae"), the ejected material helping to form interstellar clouds from which our Solar System subsequently grew. The Solar System is less than about 5 billion years old and large stars evolve to the supernova stage quite quickly so it is possible that many supernovae contributed to the material which makes up the planets. The heat released by gravitational collapse of the gaseous clouds into proto stars is sufficient (in large enough clouds) for the core to ignite a nuclear fire. Very high temperatures are required for nuclei to overcome the repulsive forces and collide with sufficient velocity to fuse. But fusion (up to iron) releases energy and so once started, the fire keeps burning. Most stars run on hydrogen fuel converting 4 hydrogen atoms (protons) into 1 helium atom (2 protons and 2 neutrons). The Sun contains enough hydrogen to produce 1056 helium nuclei and is expected to burn for about 12 billion years.

When the star exhausts its hydrogen supply, it must either step up the temperature by gravitational collapse and begin burning helium or it dies. The fate of a star depends on its size; small stars die as "white dwarfs" and large ones continue to burn successively heavier elements up to iron. Beyond iron, however, energy must be added to generate elements and we need a different mechanism for synthesizing these.

In big stars, death is the violent supernova. When the nuclear fuel is spent, the star collapses catastrophically. In the largest stars, the collapse becomes an implosion which throws off a spectacular cloud of material. It is in the supernova that elements heavier than iron are created.

The process for generating heavier elements is by "neutron capture". During stellar collapse, a burst of highly energetic neutrons is created. If a neutron collides with iron with sufficient energy, the iron nucleus will absorb the neutron. Nuclei can be built up to the size of bismuth or even larger and then undergo radioactive decay to a stable nuclide. This process of synthesis by rapid bombardment during a supernova is called the "r-process".

There are still some nuclides which can't be synthesized by the r-process. To generate these, we call on the neutrons which are generated as a by-product of normal stellar combustion. These may also be absorbed and this so called "s-process" (s for slow) accounts for most of the remaining nuclides. The few nuclides which are not explained by the mechanisms already discussed could be created by collision with protons emitted during normal stellar combustion.

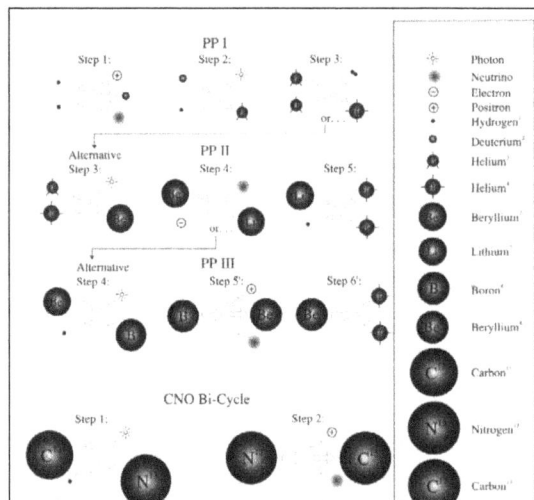

Fusion pathways.

Planetary Formation

How did the Solar System's planets come to be? The leading theory is something known as the "proto-planet hypothesis", which essentially says that very small objects stuck to each other and grew bigger and bigger—big enough to even form the gas giants, such as Jupiter.

Birthing the Sun

About 4.6 billion years ago, as the theory goes, the location of today's Solar System was nothing more than a loose collection of gas and dust—what we call a nebula.

Then something happened that triggered a pressure change in the center of the cloud, scientists say. Perhaps it was a supernova exploding nearby, or a passing star changing the gravity. Whatever the change, however, the cloud collapsed and created a disc of material, according to NASA.

The center of this disc saw a great increase in pressure that eventually was so powerful that hydrogen atoms loosely floating in the cloud began to come into contact. Eventually, they fused and produced helium, kick starting the formation of the Sun.

The Sun was a hungry youngster—it ate up 99% of what was swirling around, NASA says—but this still left 1% of the disc available for other things. And this is where planet formation began.

Time of Chaos

The Solar System was a really messy place at this time, with gas and dust and debris floating around. But planet formation appears to have happened relatively rapidly. Small bits of dust and gas began to clump together. The young Sun pushed much of the gas out to the outer Solar System and its heat evaporated any ice that was nearby.

Over time, this left rockier planets closer to the Sun and gas giants that were further away. But about four billion or so years ago, an event called the "late heavy bombardment" resulted in small bodies pelting the bigger members of the Solar System. We almost lost the Earth when a Mars-sized object crashed into it, as the theory goes.

What caused this is still under investigation, but some scientists believe it was because the gas giants were moving around and perturbing smaller bodies at the fringe of the Solar System. At any rate, in simple terms, the clumping together of proto-planets (planets in formation) eventually formed the planets.

We can still see leftovers of this process everywhere in the Solar System. There is an asteroid belt between Mars and Jupiter that perhaps would have coalesced into a planet had Jupiter's gravity not been so strong. And we also have comets and asteroids that are sometimes considered referred to as "building blocks" of our Solar System.

We've described in detail what happened in our own Solar System, but the important takeaway is that many of these processes are at work in other places. So when we speak about exoplanet systems—planets beyond our Solar System—it is believed that a similar sequence of events took place.

Making the Case

One major challenge to this theory, of course, is no one (that we know of) was recording the early history of the Solar System. That's because the Earth wasn't even formed yet, so it was impossible for any life—let alone intelligent life—to keep track of what was happening to the planets around us.

There are two major ways astronomers get around this problem:

The first is simple observation: Using powerful telescopes such as the Atacama Large Millimeter/ sub millimeter Array (ALMA), astronomers can actually observe dusty discs around young planets. So we have numerous examples of stars with planets being born around them.

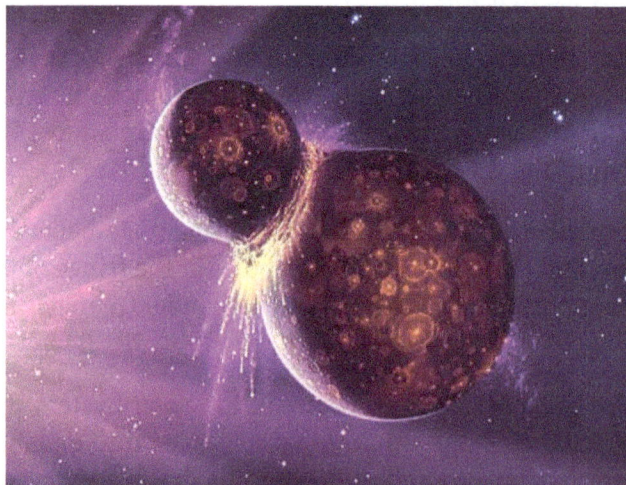

The second is using modelling: To test their observational hypotheses, astronomers run computer modelling to see if (mathematically speaking) the ideas work out. Often they will try to use different conditions during the simulation, such as perhaps a passing star triggering changes in the dust cloud. If the model holds after many runs and under several conditions, it's more likely to be true.

That said, there still are some complications. We can't use modelling yet to exactly predict how the planets of the Solar System ended up where they were. Also, in fine detail our Solar System is kind of a messy place, with phenomena such as asteroids with moons.

Models of Earth Formation

Although planets surround stars in the galaxy, how they form remains a subject of debate. Despite the wealth of worlds in our own solar system, scientists still aren't certain how planets are built. Currently, two theories are duking it out for the role of champion.

The first and most widely accepted theory, core accretion, works well with the formation of the terrestrial planets like Earth but has problems with giant planets. The second, the disk instability method, may account for the creation of these giant planets.

Scientists are continuing to study planets in and out of the solar system in an effort to better understand which of these methods is most accurate.

The Core Accretion Model

Approximately 4.6 billion years ago, the solar system was a cloud of dust and gas known as a solar nebula. Gravity collapsed the material in on itself as it began to spin, forming the sun in the center of the nebula.

With the rise of the sun, the remaining material began to clump up. Small particles drew together, bound by the force of gravity, into larger particles. The solar wind swept away lighter elements, such as hydrogen and helium, from the closer regions, leaving only heavy, rocky materials to create smaller terrestrial worlds like Earth. But farther away, the solar winds had less impact on lighter elements, allowing them to coalesce into gas giants. In this way, asteroids, comets, planets, and moons were created.

Earth's rocky core formed first, with heavy elements colliding and binding together. Dense material sank to the center, while the lighter material created the crust. The planet's magnetic field probably formed around this time. Gravity captured some of the gases that made up the planet's early atmosphere.

Early in its evolution, Earth suffered an impact by a large body that catapulted pieces of the young planet's mantle into space. Gravity caused many of these pieces to draw together and form the moon, which took up orbit around its creator.

The flow of the mantle beneath the crust causes plate tectonics, the movement of the large plates of rock on the surface of the Earth. Collisions and friction gave rise to mountains and volcanoes, which began to spew gases into the atmosphere.

Although the population of comets and asteroids passing through the inner solar system is sparse today, they were more abundant when the planets and sun were young. Collisions from these icy bodies likely deposited much of the Earth's water on its surface. Because the planet is in the Goldilocks zone, the region where liquid water neither freezes nor evaporates but can remain as a liquid, the water remained at the surface, which many scientists think plays a key role in the development of life.

Exoplanet observations seem to confirm core accretion as the dominant formation process. Stars with more "metals" — a term astronomers use for elements other than hydrogen and helium — in their cores have more giant planets than their metal-poor cousins. According to NASA, core accretion suggests that small, rocky worlds should be more common than the more massive gas giants.

The 2005 discovery of a giant planet with a massive core orbiting the sun-like star HD 149026 is an example of an exoplanet that helped strengthen the case for core accretion.

"This is a confirmation of the core accretion theory for planet formation and evidence that planets of this kind should exist in abundance," said Greg Henry in a press release. Henry, an astronomer at Tennessee State University, Nashville, detected the dimming of the star.

In 2017, the European Space Agency plans to launch the characterising exoplanet Satellite (CHEOPS), which will study exoplanets ranging in sizes from super-Earths to Neptune. Studying these distant worlds may help determine how planets in the solar system formed.

"In the core accretion scenario, the core of a planet must reach a critical mass before it is able to accrete gas in a runaway fashion," said the CHEOPS team.

"This critical mass depends upon many physical variables, among the most important of which is the rate of planetesimals accretion."

By studying how growing planets accrete material, cheops will provide insight into how worlds grow.

The Disk Instability Model

Although the core accretion model works fine for terrestrial planets, gas giants would have needed to evolve rapidly to grab hold of the significant mass of lighter gases they contain. But simulations have not been able to account for this rapid formation. According to models, the process takes several million years, longer than the light gases were available in the early solar system. At the same time, the core accretion model faces a migration issue, as the baby planets are likely to spiral into the sun in a short amount of time.

According to a relatively new theory, disk instability, clumps of dust and gas are bound together early in the life of the solar system. Over time, these clumps slowly compact into a giant planet. These planets can form faster than their core accretion rivals, sometimes in as little as a thousand years, allowing them to trap the rapidly-vanishing lighter gases. They also quickly reach an orbit-stabilizing mass that keeps them from death-marching into the sun.

According to exoplanetary astronomer Paul Wilson, if disk instability dominates the formation of planets, it should produce a wide number of world's at large orders. The four giant planets orbiting at significant distances around the star HD 9799 provides observational evidence for disk instability. Fomalhautb, an exoplanet with a 2,000-year orbit around its star, could also be an example of a world formed through disk instability, though the planet could also have been ejected due to interactions with its neighbors.

Pebble Accretion

The biggest challenge to core accretion is time — building massive gas giants fast enough to grab the lighter components of their atmosphere. Recent research on how smaller, pebble-sized objects fused together to build giant planets up to 1000 times faster than earlier studies.

"This is the first model that we know about that you start out with a pretty simple structure for the solar nebula from which planets form, and end up with the giant-planet system that we see," study lead author Harold Levison, an astronomer at the Southwest Research Institute (SwRI) in Colorado.

In 2012, researchers Michiel Lambrechts and Anders Johansen from Lund University in Sweden proposed that tiny pebbles, once written off, held the key to rapidly building giant planets.

"They showed that the leftover pebbles from this formation process, which previously were thought to be unimportant, could actually be a huge solution to the planet-forming problem," Levison said.

Levison and his team built on that research to model more precisely how the tiny pebbles could form planets seen in the galaxy today. While previous simulations, both large and medium-sized objects consumed their pebble-sized cousins at a relatively constant rate, Levison's simulations suggest that the larger objects acted more like bullies, snatching away pebbles from the mid-sized masses to grow at a far faster rate.

"The larger objects now tend to scatter the smaller ones more than the smaller ones scatter them back, so the smaller ones end up getting scattered out of the pebble disk," study co-author Katherine Kretke, also from SwRI, told Space.com. "The bigger guy basically bullies the smaller one so they can eat all the pebbles themselves, and they can continue to grow up to form the cores of the giant planets."

Evolution of the Earth

The blue, cloud-enveloped planet the we recognize immediately from satellite pictures seems remarkably stable. Continents and oceans, encircled by an oxygen-rich atmosphere, support familiar life-forms. Yet this constancy is an illusion produced by the human experience of time. Earth and its atmosphere are continuously altered. Plate tectonics shift the continents, raise mountains and move the ocean floor while processes not fully understood alter the climate.

Such constant change has characterized Earth since its beginning some 4.5 billion years ago. From the outset, heat and gravity shaped the evolution of the planet. These forces were gradually joined by the global effects of the emergence of life. Exploring this past offers us the only possibility of understanding the origin of life and, perhaps, its future.

Scientists used to believe the rocky planets, including Earth, Mercury, Venus and Mars, were created by the rapid gravitational collapse of a dust cloud, a deation giving rise to a dense orb. In the 1960s the Apollo space program changed this view. Studies of moon craters revealed that these gouges were caused by the impact of objects that were in great abundance about 4.5 billion years ago. Thereafter, the number of impacts appeared to have quickly decreased. This observation rejuvenated the theory of accretion postulated by Otto Schmidt. The Russian geophysicist had suggested in 1944 that planets grew in size gradually, step by step.

According to Schmidt, cosmic dust lumped together to form particulates, particulates became gravel, gravel became small balls, then big balls, then tiny planets, or planetesimals, and, nally, dust became the size of the moon. As the planetesimals became larger, their numbers decreased. Consequently, the number of collisions between planetesimals, or meteorites, decreased. Fewer items available for accretion meant that it took a long time to build up a large planet. A calculation made by George W. Wetherill of the Carnegie Institution of Washington suggests that about 100 million years could pass between the formation of an object measuring 10 kilometers in diameter and an object the size of Earth.

The process of accretion had significant thermal consequences for Earth, consequences that forcefully directed its evolution. Large bodies slamming into the planet produced immense heat in its interior, melting the cosmic dust found there. The resulting furnace situated some 200 to 400 kilometers underground and called a magma ocean was active for millions of years, giving rise to volcanic eruptions. When Earth was young, heat at the surface caused by volcanism and lava ows from the interior was intensified by the constant bombardment of huge objects, some of them perhaps the size of the moon or even Mars. No life was possible during this period.

Beyond clarifying that Earth had formed through accretion, the Apollo program compelled scientists to try to reconstruct the subsequent temporal and physical development of the early Earth. This undertaking had been considered impossible by founders of geology, including Charles Lyell, to whom the following phrase is attributed: No vestige of a beginning, no prospect for an end. This statement conveys the idea that the young Earth could not be re-created, because its remnants were destroyed by its very activity. But the development of isotope geology in the 1960s had rendered this view obsolete. Their imaginations red by Apollo and the moon endings, geochemists began to apply this technique to understand the evolution of Earth.

Dating rocks using so-called radioactive clocks allows geologists to work on old terrains that do not contain fossils. The hands of a radioactive clock are isotopes-atoms of the same element that have different atomic weights-and geologic time is measured by the rate of decay of one isotope into another. Among the many clocks, those based on the decay of uranium 238 into lead 206 and of uranium 235 into lead 207 are special. Geochronologists can determine the age of samples by analyzing only the daughter product-in this case, lead-of the radioactive parent, uranium.

Panning for Zircons

Isotope Geology has permitted geologists to determine that the accretion of Earth culminated in the differentiation of the planet: the creation of the core the source of Earth's magnetic field-and the beginning of the atmosphere. In 1953 the classic work of Claire C. Patterson of the California Institute of Technology used the uranium-lead clock to establish an age of 4.55 billion years for Earth and many of the meteorites that formed it.

As Patterson argued, some meteorites were indeed formed about 4.56 billion years ago, and their debris constituted Earth. But Earth continued to grow through the bombardment of planetesimals until some 120 million to 150 million years later. At that time 4.44 billion to 4.41 billion years ago Earth began to retain its atmosphere and create its core. This possibility had already been suggested by Bruce R. Doe and Robert E. Zartman of the U.S. Geological Survey in Denver two decades ago and is in agreement with Wetherills estimates.

The emergence of the continents came somewhat later. According to the theory of plate tectonics, these landmasses are the only part of Earth's crust that is not recycled and, consequently, destroyed during the geothermal cycle driven by the convection in the mantle. Continents thus provide a form of memory because the record of early life can be read in their rocks. Geologic activity, however, including plate tectonics, erosion and metamorphism, has destroyed almost all the ancient rocks. Very few fragments have survived this geologic machine.

Nevertheless, in recent decades, several important ends have been made, again using isotope geochemistry. One group, led by Stephen Moorbath of the University of Oxford, discovered terrain in West Greenland that is between 3.7 billion and 3.8 billion years old. In addition, Samuel A. Bowring of the Massachusetts Institute of Technology explored a small area in North America the Acasta gneiss that is thought to be 3.96 billion years old.

Ultimately, a quest for the mineral zircon led other researchers to even more ancient terrain. Typically found in continental rocks, zircon is not dissolved during the process of erosion but is deposited in particle form in sediment. A few pieces of zircon can therefore survive for billions of years and can serve as a witness to Earths more ancient crust. The search for old zircons started in Paris with the work of Annie Vitrac and Jol R. Lancelot, later at the University of Marseille and now at the University of Nimes, respectively, as well as with the efforts of Moorbath and Allgre. It was a group at the Australian National University in Canberra, directed by William Compston that was nally successful. The team discovered zircons in western Australia that were between 4.1 billion and 4.3 billion years old.

Zircons have been crucial not only for understanding the age of the continents but for determining when life rst appeared. The earliest fossils of undisputed age were found in Australia and South Africa. These relics of blue-green algae are about 3.5 billion years old. Manfred Schidlowski of the Max Planck Institute for Chemistry in Mainz studied the Issue formation in West Greenland and argued that organic matter existed as long ago as 3.8 billion years. Because most of the record of early life has been destroyed by geologic activity, we cannot say exactly when it rst appeared-perhaps it arose very quickly, maybe even 4.2 billion years ago.

Stories from Gases

One of the most important aspects of the planet's evolution is the formation of the atmosphere, because it is this assemblage of gases that allowed life to crawl out of the oceans and to be sustained. Researchers have hypothesized since the 1950s that the terrestrial atmosphere was created by gases emerging from the interior of the planet. When a volcano spews gases, it is an example of the continuous outgassing, as it is called, of Earth. But scientists have questioned whether this process occurred suddenly-about 4.4 billion years ago when the core differentiated-or whether it took place gradually over time.

To answer this question, Allègre and his colleagues studied the isotopes of rare gases. These gases-including helium, argon and xenon-have the peculiarity of being chemically inert, that is, they do not react in nature with other elements. Two of them are particularly important for atmospheric studies: argon and xenon. Argon has three isotopes, of which argon 40 is created by the decay of potassium 40. Xenon has nine, of which xenon 129 has two different origins. Xenon 129 arose as the result of nucleo synthesis before Earth and solar system were formed. It was also created from the decay of radioactive iodine 129, which does not exist on Earth anymore. This form of iodine was present very early on but has died out since, and xenon 129 has grown at its expense.

Like most couples, both argon 40 and potassium 40 and xenon 129 and iodine 129 have stories to tell. They are excellent chronometers. Although the atmosphere was formed by the outgassing of the mantle, it does not contain any potassium 40 or iodine 129. All argon 40 and xenon 129, formed in Earth and released, are found in the atmosphere today. Xenon was expelled from the mantle and retained in the atmosphere; therefore, the atmosphere-mantle ratio of this element allows us to evaluate the age of differentiation. Argon and xenon trapped in the mantle evolved by the radioactive decay of potassium 40 and iodine 129. Thus, if the total outgassing of the mantle occurred at the beginning of Earths formation, the atmosphere would not contain any argon 40 but would contain xenon 129.

The major challenge facing an investigator who wants to measure such ratios of decay is to obtain high concentrations of rare gases in mantle rocks because they are extremely limited. Fortunately, a natural phenomenon occurs at mid-ocean ridges during which volcanic lava transfers some silicates from the mantle to the surface. The small amounts of gases trapped in mantle minerals rise with the melt to the surface and are concentrated in small vesicles in the outer glassy margin of lava ows. This process serves to concentrate the amounts of mantle gases by a factor of 104 or 105. Collecting these rocks by dredging the seaoor and then crushing them under vacuum in a sensitive mass spectrometer allows geochemists to determine the ratios of the isotopes in the mantle. The results are quite surprising. Calculations of the ratios indicate that between 80 and 85 percent of the atmosphere was outgassed during Earths rst one million years; the rest was released slowly but constantly during the next 4.4 billion years.

The composition of this primitive atmosphere was most certainly dominated by carbon dioxide, with nitrogen as the second most abundant gas. Trace amounts of methane, ammonia, sulfur dioxide and hydrochloric acid were also present, but there was no oxygen. Except for the presence of abundant water, the atmosphere was similar to that of Venus or Mars. The details of the evolution of the original atmosphere are debated, particularly because we do not know how strong the sun was at that time. Some facts, however, are not disputed. It is evident that carbon dioxide played a crucial role. In addition, many scientists believe the evolving atmosphere contained sufficient quantities of gases such as ammonia and methane to give rise to organic matter.

Still, the problem of the sun remains unresolved. One hypothesis holds that during the Archean eon, which lasted from about 4.5 billion to 2.5 billion years ago, the suns power was only 75 percent of what it is today. This possibility raises a dilemma: How could life have survived in the relatively cold climate that should accompany a weaker sun? A solution to the faint early sun paradox, as it is called, was offered by Carl Sagan and George Mullen of Cornell University in 1970. The two scientists suggested that methane and ammonia, which are very effective at trapping infrared radiation, were quite abundant. These gases could have created a super-greenhouse effect. The idea was criticized on the basis that such gases were highly reactive and have short lifetimes in the atmosphere.

Controlled C. O.

In the late 1970s Veerabhadran Ramanathan, now at the Scripps Institution of Oceanography, and Robert D. Cess and Tobias Owen of Stony Brook University proposed another solution. They postulated that there was no need for methane in the early atmosphere because carbon dioxide was abundant enough to bring about the super-greenhouse effect. Again this argument raised a different question: How much carbon dioxide was there in the early atmosphere? Terrestrial carbon dioxide is now buried in carbonate rocks, such as limestone, although it is not clear when it became trapped there. Today calcium carbonate is created primarily during biological activity; in the Archean eon, carbon may have been primarily removed during inorganic reactions.

The rapid outgassing of the planet liberated voluminous quantities of water from the mantle, creating the oceans and the hydrologic cycle. The acids that were probably present in the atmosphere eroded rocks, forming carbonate-rich rocks. The relative importance of such a mechanism is, however, debated. Heinrich D. Holland of Harvard University believes the amount of carbon dioxide in the atmosphere rapidly decreased during the Archean and stayed at a low level.

Understanding the carbon dioxide content of the early atmosphere is pivotal to understanding climatic control. Two conicting camps have put forth ideas on how this process works. The rst group holds that global temperatures and carbon dioxide were controlled by inorganic geochemical feedbacks; the second asserts that they were controlled by biological removal.

James C. G. Walker, James F. Kasting and Paul B. Hays, then at the University of Michigan at Ann Arbor, proposed the inorganic model in 1981. They postulated that levels of the gas were high at the outset of the Achaean and did not fall precipitously. The trio suggested that as the climate warmed, more water evaporated, and the hydrologic cycle became more vigorous, increasing precipitation and runoff. The carbon dioxide in the atmosphere mixed with rainwater to create carbonic acid runoff, exposing minerals at the surface to weathering. Silicate minerals combined with carbon that had been in the atmosphere, sequestering it in sedimentary rocks. Less carbon dioxide in the atmosphere meant, in turn, less of a greenhouse effect. The inorganic negative feedback process offset the increase in solar energy.

This solution contrasts with a second paradigm: biological removal. One theory advanced by James E. Lovelock, an originator of the Gaia hypothesis, assumed that photosynthesizing microorganisms, such as phytoplankton, would be very productive in a high carbon dioxide environment. These creatures slowly removed carbon dioxide from the air and oceans, converting it into calcium carbonate sediments. Critics retorted that phytoplankton had not even evolved for most of the time that Earth has had life. (The Gaia hypothesis holds that life on Earth has the capacity to regulate temperature and the composition of Earth's surface and to keep it comfortable for living organisms).

In the early 1990s Tyler Volk of New York University and David W. Schwartzman of Howard University proposed another Gaian solution. They noted that bacteria increase carbon dioxide content in soils by breaking down organic matter and by generating humic acids. Both activities accelerate weathering, removing carbon dioxide from the atmosphere. On this point, however, the controversy becomes acute. Some geochemists, including Kasting, now at Pennsylvania State University, and Holland, postulate that while life may account for some carbon dioxide removal after the Archean, inorganic geochemical processes can explain most of the sequestering. These researchers view life as a rather weak climatic stabilizing mechanism for the bulk of geologic time.

Oxygen from Algae

The issue of carbon remains critical to how life intended the atmosphere. Carbon burial is a key to the vital process of building up atmospheric oxygen concentrations-a prerequisite for the development of certain life-forms. In addition, global warming is taking place now as a result of humans releasing this carbon. For one billion or two billion years, algae in the oceans produced oxygen. But because this gas is highly reactive and because there were many reduced minerals in the ancient oceans-iron, for example, is easily oxidized-much of the oxygen produced by living creatures simply got used up before it could reach the atmosphere, where it would have encountered gases that would react with it.

Even if evolutionary processes had given rise to more complicated life-forms during this anaerobic era, they would have had no oxygen. Furthermore, un ltered ultraviolet sunlight would have likely killed them if they left the ocean. Researchers such as Walker and Preston Cloud, then at the University of California at Santa Barbara, have suggested that only about two billion years ago, after most of the reduced minerals in the sea were oxidized, did atmospheric oxygen accumulate. Between one billion and two billion years ago oxygen reached current levels, creating a niche for evolving life.

By examining the stability of certain minerals, such as iron oxide or uranium oxide, Holland has shown that the oxygen content of the Archean atmosphere was low before two billion years ago. It is largely agreed that the present-day oxygen content of 20 percent is the result of photosynthetic activity. Still, the question is whether the oxygen content in the atmosphere increased gradually over time or suddenly. Recent studies indicate that the increase of oxygen started abruptly between 2.1 billion and 2.03 billion years ago and that the present situation was reached 1.5 billion years ago.

The presence of oxygen in the atmosphere had another major benefit for an organism trying to live at or above the surface: it littered ultraviolet radiation. Ultraviolet radiation breaks down many molecules-from DNA and oxygen to the chloroform carbons that are implicated in stratospheric ozone depletion. Such energy splits oxygen into the highly unstable atomic form O, which can combine back into O_2 and into the very special molecule O_3, or ozone. Ozone, in turn, absorbs ultraviolet radiation. It was not until oxygen was abundant enough in the atmosphere to allow the formation of ozone that life even had a chance to get a root-hold or a foothold on land. It is not a coincidence that the rapid evolution of life from prokaryotes (single-celled organisms with no nucleus) to eukaryotes (single-celled organisms with a nucleus) to metazoan (multi-celled organisms) took place in the billion-year-long era of oxygen and ozone.

Although the atmosphere was reaching a fairly stable level of oxygen during this period, the climate was hardly uniform. There were long stages of relative warmth or coolness during the transition to modern geologic time. The composition of fossil plankton shells that lived near the ocean provides a measure of bottom water temperatures. The record suggests that over the past 100 million years bottom waters cooled by nearly 15 degrees Celsius. Sea levels dropped by hundreds of meters, and continents drifted apart. Inland seas mostly disappeared, and the climate cooled an average of 10 to 15 degrees C. Roughly 20 million years ago permanent ice appears to have built up on Antarctica.

About two million to three million years ago the pale climatic record starts to show significant expansions and contractions of warm and cold periods in 40,000-year or so cycles. This periodicity is interesting because it corresponds to the time it takes Earth to complete an oscillation of the tilt of its axis of rotation. It has long been speculated, and recently calculated, that known changes in orbital geometry could alter the amount of sunlight coming in between winter and summer by about 10 percent or so and could be responsible for initiating or ending ice ages.

The Warm Hand of Man

Most interesting and perplexing is the discovery that between 600,000 and 800,000 years ago the dominant cycle switched from 40,000-year periods to 100,000-year intervals with very large actuations. The last major phase of glaciation ended about 10,000 years ago. At its height 20,000 years ago, ice sheets about two kilometers thick covered much of northern Europe and North America. Glaciers expanded in high plateaus and mountains throughout the world. Enough ice was locked up on land to cause sea levels to drop more than 100 meters below where they are today. Massive ice sheets scoured the land and revamped the ecological face of Earth, which was +ve degrees C cooler on average than it is currently.

The precise causes of the longer intervals between warm and cold periods are not yet sorted out. Volcanic eruptions may have played a significant role, as shown by the effect of El Chi chon in Mexico and Mount Pinatubo in the Philippines. Tectonic events, such as the development of the Himalayas, may have influenced world climate. Even the impact of comets can influence short-term climatic trends with catastrophic consequences for life. It is remarkable that despite violent, episodic perturbations, the climate has been buffered enough to sustain life for 3.5 billion years.

One of the most pivotal climatic discoveries of the past 30 years has come from ice cores in Greenland and Antarctica. When snow falls on these frozen continents, the air between the snow grains is trapped as bubbles. The snow is gradually compressed into ice, along with its captured gases. Some of these records can go back more than 500,000 years; scientists can analyze the chemical content of ice and bubbles from sections of ice that lie as deep as 3,600 meters (2.2 miles) below the surface.

The ice-core borers have determined that the air breathed by ancient Egyptians and Anasazi Indians was very similar to that which we inhale today-except for a host of air pollutants introduced over the past 100 or 200 years. Principal among these added gases, or pollutants, are extra carbon dioxide and methane. Since about 1860-the expansion of the Industrial Revolution-carbon dioxide levels in the atmosphere have increased more than 30 percent as a result of industrialization and deforestation; methane levels have more than doubled because of agriculture, land use and energy production. The ability of increased amounts of these gases to trap heat is what drives concerns about climate change in the 21st century.

The ice cores have shown that sustained natural rates of worldwide temperature change are typically about one degree C per millennium. These shifts are still significant enough to have radically altered where species live and to have potentially contributed to the extinction of such charismatic mega fauna as mammoths and saber-toothed tigers. But a most extraordinary story from the ice cores is not the relative stability of the climate during the past 10,000 years. It appears that during the height of the last ice age 20,000 years ago there was 50 percent less carbon dioxide and less than half as much methane in the air than there has been during our epoch, the Holocene. This suggests a positive feedback between carbon dioxide, methane and climatic change.

The reasoning that supports the idea of this destabilizing feedback system goes as follows. When the world was colder, there was less concentration of greenhouse gases, and so less heat was trapped. As Earth warmed up, carbon dioxide and methane levels increased, accelerating the warming. If life had a hand in this story, it would have been to drive, rather than to oppose, climatic change. It appears increasingly likely that when humans became part of this cycle, they, too, helped to accelerate warming. Such warming has been especially pronounced since the mid-1800s because of greenhouse gas emissions from industrialization, land-use change and other phenomena. Once again, though, uncertainties remain.

Nevertheless, most scientists would agree that life could well be the principal factor in the positive feedback between climatic change and greenhouse gases. There was a rapid rise in average global surface temperature at the end of the 20th century. Indeed, the period from the 1980s onward has been the warmest of the past 2,000 years. Nineteen of the 20 warmest years on record have occurred since 1980, and the 12 warmest have all occurred since 1990. The all-time record high year was 1998, and 2002 and 2003 were in second and third places, respectively. There is good reason to believe that the decade of the 1990s would have been even hotter had not Mount Pinatubo erupted: this volcano put enough dust into the high atmosphere to block some incident sunlight, causing global cooling of a few tenths of a degree for several years.

The box at the right shows a remarkable study that attempted to push back the Northern Hemisphere's temperature record a full 1,000 years. Climatologist Michael Mann of the University of Virginia and his colleagues performed a complex statistical analysis involving some 112 different factors related to temperature, including tree rings, the extent of mountain glaciers, changes in coral reefs, sunspot activity and volcanism.

The resulting temperature record is a reconstruction of what might have been obtained had thermometer-based measurements been available. As shown by the confidence range, there is considerable uncertainty in each year of this 1,000-year temperature reconstruction. But the overall trend is clear: a gradual temperature decrease over the first 900 years, followed by a sharp temperature upturn in the 20th century. This graph suggests that the decade of the 1990s was not only the warmest of the century but of the entire past millennium.

By studying the transition from the high carbon dioxide, low-oxygen atmosphere of the Archean to the era of great evolutionary progress about half a billion years ago, it becomes clear that life may have been a factor in the stabilization of climate. In another example-during the ice ages and interglacial cycles-life seems to have the opposite function: accelerating the change rather than diminishing it. This observation has led one of us (Schneider) to contend that climate and life coevolved rather than life serving solely as a negative feedback on climate.

If we humans consider ourselves part of life-that is, part of the natural system-then it could be argued that our collective impact on Earth means we may have a significant co-evolutionary role in the future of the planet. The current trends of population growth, the demands for increased standards of living and the use of technology and organizations to attain these growth-oriented goals all contribute to pollution. When the price of polluting is low and the atmosphere is used as a free sewer, carbon dioxide, methane, chlorouoro carbons, nitrous oxides, sulfur oxides and other toxics can build up.

Drastic Changes Ahead

In their report Climate Change 2001, climate experts on the Intergovernmental Panel on Climate Change estimated that the world will warm between 1.4 and 5.8 degrees C by 2100. The mild end of that range-a warming rate of 1.4 degrees C per 100 years-is still 14 times faster than the one degree C per 1,000 years that historically has been the average rate of natural change on a global scale. Should the higher end of the range occur, then we could see rates of climatic change nearly 60 times faster than natural average conditions, which could lead to changes that many would consider dangerous. Change at this rate would almost certainly force many species to attempt to move their ranges, just as they did from the ice age/interglacial transition between 10,000 and 15,000 years ago. Not only would species have to respond to climatic change at rates 14 to 60 times faster, but few would have undisturbed, open migration routes as they did at the end of the ice age and the onset of the interglacial era. The negative effects of this significant warming-on health, agriculture, coastal geography and heritage sites, to name a few-could also be severe.

To make the critical projections of future climatic change needed to understand the fate of ecosystems on Earth, we must dig through land, sea and ice to learn as much from geologic, pale climatic and pale ecological records as we can. These records provide the backdrop against which to calibrate the crude instruments we must use to peer into a shadowy environmental future, a future increasingly influenced by us.

Composition of the Earth

Sources of Information

Our knowledge of earth's internal characteristics of these concentric layers has been acquired through direct as well as indirect sources.

It is true that most of our knowledge about the interior of the earth is largely based on estimates and inferences.

Since we are unable to drill more than a few kilometers into the earth, our information about various concentric layers have been acquired entirely through indirect evidences.

However, a part of the information is gathered through direct observations and analysis of material. Thus, there have been (i) direct sources, and (ii) indirect sources of our knowledge about the interior of the earth.

Direct Sources

We get a lot of easily available solid geometries in surface rock or the rocks procured from the mines. But through mining and drilling scientists have gained a very limited knowledge of the earth's interior.

As we know, the earth has a radius of about 6400 kilometers, but man has succeeded to penetrate and examine directly only its outer skin.

Thus, through mining and drilling we have gained a very scanty knowledge of the earth's interior. Gold mines in South Africa are as deep as 3-4 km. going beyond this depth is impossible because of very intense heat at this depth.

The rate of increase in temperature with depth which is known as the geothermal gradient has an average value of about 30°C per kilometer.

Besides, oil exploration has penetrated to about 7.2 km by drilling. These explorations could provide knowledge only about the earth's uppermost layers.

The second direct source though which direct information is obtained has been the volcanic eruption throwing onto the surface of the earth the molten rock material called magma. But it has not been possible to determine the exact depth of its origin.

Indirect Sources

Through the mining activities, it became clear that temperature and pressure increase with increasing depth of the mines. It is also known that there is an increase in the density of the material with depth.

However, the rate of change of these characteristics i.e. temperature, pressure and density can the calculated. Knowing the total thickness of the earth, the values of above characteristics at different depths have been estimated by scientists.

Meteors may be considered as another source of information. However, these materials available for analysis do not come from the earth's interior.

But there is little doubt that material and the structure observed in the meteors are more or less similar to that of the earth because they are solid bodies developed out of materials similar to our planet. So it may be considered another source of information about the interior of the earth.

Gravitation, magnetic field and seismic activity are other indirect sources of knowledge about the interior of the earth. Gravity anomalies found at different places on the earth's surface give us information regarding the unequal distribution of mass of material in the crust of the earth.

In the same way, magnetic surveys also give information about the distribution of magnetic materials in the earth's crust. Seismic activity is undoubtedly the most valuable source of information about the interior of the earth.

The Zonal Structure of the Earth based on Earthquake Waves

Because of tectonic processes occurring in various parts of the Earth, elastic energy is accumulated in the rocks. When this energy attains the breaking strain of the particular rock, an earthquake occurs and the energy propagates in the form of elastic waves. Based on observations the earthquakes do not occur at depths exceeding 700 km. Consequently, elastic energy is not accumulated or cannot be accumulated in materials deeper than that. Consequently, material at depths exceeding 700 km either has plastically behavior against durable and not too great deformations or it is eventually in a state of complete rest. (If the latter were true, it would lead to contradictions.) According to the solution of the wave equation describing the propagation of elastic energy, two types of spatial waves are able to propagate in a solid elastic medium: longitudinal or dilatational waves (P) and transversal or shear waves (S). The velocities of these two waves are expressed by the formula.

$$v_P = \sqrt{\frac{K + \frac{4}{3}\mu}{\vartheta}}$$

and

$$v_s = \sqrt{\frac{\mu}{\vartheta}}$$

Where J is the density of the medium, K is the bulk modulus and m is the shear modulus. It follows from relation ($v_s = \sqrt{\frac{\mu}{\vartheta}}$) that since the rigidity of liquids is zero ($= \mu 0$), $0 = sv$ as well, i.e. the shear waves do not propagate in liquids. In a homogeneous isotropic medium, the elastic waves starting from one point propagate in a homogeneous isotropic way. When, however, they reach the boundary of two different media they are reflected, and they creating each other according to the Descartes' law. Only longitudinal waves arrive at the given boundary, in addition to these longitudinal waves, transversal waves will be reflected from the boundary and generate into the other medium.

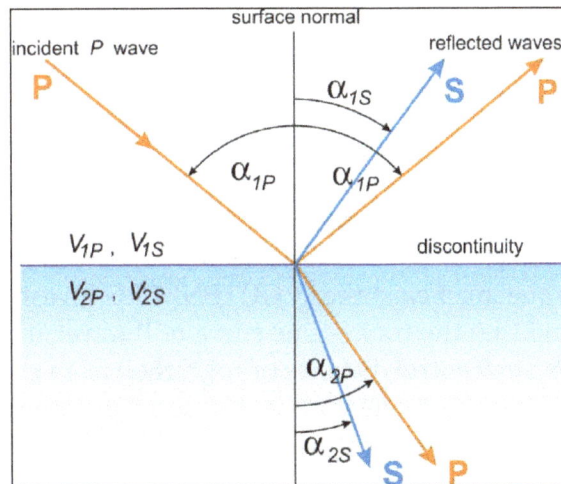

Reflection and refraction of seismic waves.

When in the hypocenter of an earthquake a quake of adequate energy occurs, the arrival of the seismic waves will be recorded after a certain time in the seismological stations located at various points of the Earth. Elastic waves arising from greater epicentral distances bring information from the deeper parts of the Earth.

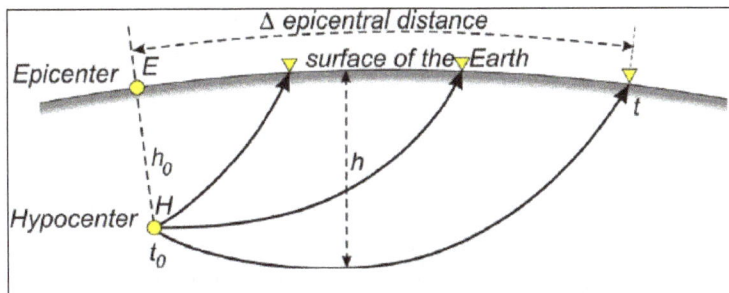

Propagation of seismic waves from the hypocenter.

Since the velocity of seismic waves according to relations $v_P = \sqrt{\dfrac{K + \dfrac{4}{3}\mu}{\vartheta}}$ and $v_s = \sqrt{\dfrac{\mu}{\vartheta}}$ has a different

value for the various materials, it is possible to draw conclusions on the velocities from the time of arrival of these waves and thus in an indirect way also on the physical properties (eventually the material composition) of the medium through which the waves passed. Analysis of seismic waves has shown that discontinuities exist inside the Earth and that the Earth is divided by these discontinuities into zones having different physical parameters.

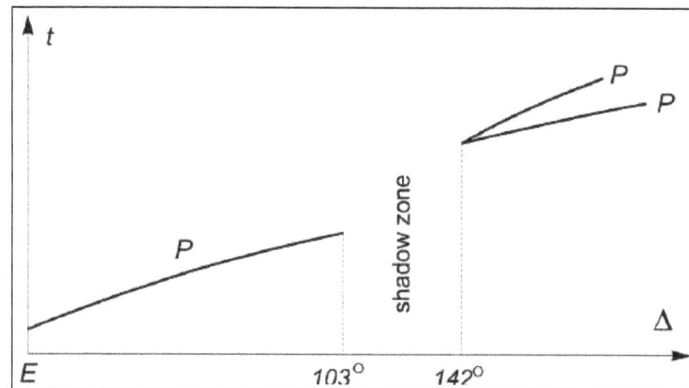

Schematic travel-time curves for waves passing through a two-layered sphere if the velocity increases.

The initial observations pointed out that the velocity of seismic waves increases with depth. However, R.D. Oldham called attention, as far back as 1906, to the fact that the P waves arrive much later than the expected time at the seismological stations located on the side of the Earth opposite the epicenter of the earthquake. Thus, since these waves pass through the central parts of the Earth, a lower velocity zone must exist there. GUTENBERG investigated this phenomenon in some detail in 1914. He found that the travel-time curve of P waves increases steadily in the way shown in figure above 0° to an epicentral distance of 103°, then from 103° to 142° the longitudinal waves are completely absent whereas from 142° on, the curve is decomposed to two parts in that immediately after the first P wave another P wave arrives to the surface. This so-called shadow zone between 103° and 142° is due, according to GUTENBERG's calculations, to the existence, in the inside of the Earth at a depth of 2900 km, of a boundary which, when it is passed results in an abrupt decrease of the velocity of earthquake waves. By this so-called Gutenberg Wincher boundary, the inside of the Earth is divided into two parts: the mantle and the core.

the quake sourced at the hypocenter denoted by H. The quake waves cover, with increasing epicentral distance, an ever-longer way, and they are immersed ever deeper into the mantle. The wave arriving at an epicentral distance of 103° may just pass beside the core but the next wave collides with the core. Thus, since the seismic velocity is lower in the core, the wave is refracted towards the incident perpendicular then, on attaining again the boundary of the core after passing through the core, it is refracted from the incident perpendicular, arriving at the surface at a distance of about 190° from the epicenter of the quake. The next waves whose incident angles at the boundary of the core become smaller and smaller cover a similar way and emerge at the surface at smaller and smaller epicentral distances. The smallest epicentral distance is 142°. If the incident angle of the quake waves decreases further, the waves attain the surface of the Earth at ever-increasing epicentral distances.

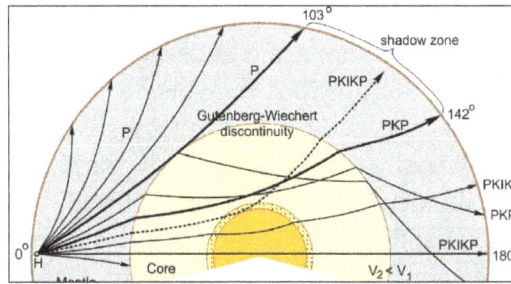

Vertical section through half the Earth, showing the propagation of the longitudinal waves from a hypocenter (transition zone between outer and inner core is dotted).

According to further detailed investigations of lechman, even the shadow zone is not completely free of the P waves some weak longitudinal waves can be recorded even here. From this fact, he concluded that even the core itself is not homogeneous and it may be divided into an outer and an inner core. In the inner core, the velocity of P waves is much higher than that in the exterior part of the core and thus the waves arriving at an adequate angle are refracted in a way that they attain the surface of the Earth within the shadow zone. It is rather difficult to determine the boundary between the outer and the inner core since this boundary is not as sharp as the separating line between the core and the mantle. Here instead of the boundary a transition layer whose thickness is about 100 km, is presumed by seismologists . This so-called Lehman zone separates the outer and the inner core at a depth of about 5000 to 5100 km. According to BARTA the uncertainty of the depth of the Lehman zone may also be due to the fact that the inner core is not located exactly in the geometrical centre of the Earth[6, 7, 8], and thus calculations from the waves of earthquakes occurring at various points of the Earth give different depths.

The outermost and best-known zone of the Earth is its crust, which can by no means be considered as a homogeneous zone. In 1909 A. MOHOROVICIC, the Croatian geo-physicist, was the first to indicate that under the Balkan peninsula, at a depth of about 50 km, a boundary can be found below which a rapid velocity increase can be observed . Seismological investigations carried out later proved that this boundary could be found throughout the Earth at an average depth of 33 km. This boundary named after its discoverer as the Mohorovicic (abbreviated to Moho) discontinuity can be considered as the lower boundary of the Earth's crust representing the boundary between the crust and the mantle. Thus it can be stated that in the inside of the Earth two boundaries exist on passing through which the seismic velocities are altered rapidly. These two boundaries are de-noted as the Moho and Gutenberg-Wiechert discontinuities and they divide the Earth into three main zones: the crust, the mantle and the core. However, the seismic velocities are not constant in any of the mentioned three zones but varies.

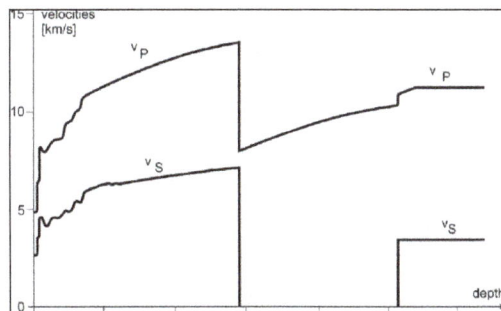

Variation of seismic velocities in the Earth's interior.

The velocity of seismic waves (P and S waves) advancing to the inside of the Earth can be determined at various depths in the knowledge of the apparent superficial velocity and of the depth of immersion of the waves. It can be seen in figure that the seismic waves emerging at ever-increasing epicentral distances immerse into ever-deeper parts of the Earth and supply information concerning the conditions of velocity existing there. Based on a variation the three main zones of the Earth mentioned earlier can be divided into further sub-units. Namely, though the velocities increase steadily within the same main zone, the measure of the increase (the first derivative of the function) changes abruptly at certain depths. Spots where the velocity of seismic waves, abruptly changes in the way are denoted as surfaces of discontinuities of first order whereas those where their derivatives change as those of second order. In the plot of the variation of seismic velocities with depth the great velocity decrease at a depth of 2900 km is striking. In the mantle of the Earth, the velocity of the longitudinal and transversal waves shows a gradual increase. At the bottom of the mantle, the velocity of the P waves attains 13.6 km/s whereas that of the S waves is 7.3 km/s. On passing through the mantle-core boundary, the velocity of the longitudinal waves decreases abruptly to 8.1 km/s then later gradually increases in the Lehman zone (interpreted in different ways by the various researchers) and attains in the centre of the Earth a value of 11.1 km/s. It is most striking that the velocity of propagation of the transversal waves decreases to zero at the boundary of the core. According to relation $v_s = \sqrt{\dfrac{\mu}{\vartheta}}$ this is possible only when $o = \mu$, that is, the shear modulus zero. However, from this it follows that the outer core behaves as liquid.

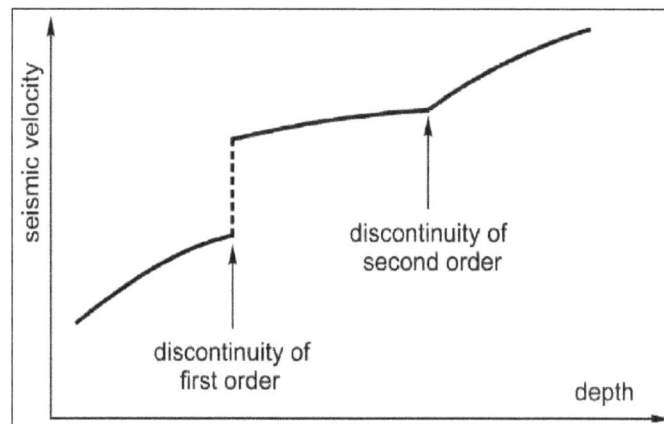

Discontinuities of first and second order.

In the knowledge of P and S velocity distributions the zonal structure of the Earth can be determined quite accurately. The zonal structure showing the depths of the discontinuities of first and second order, the denotations of the individual discontinuities and zones, furthermore the relative volumes of the individual zones can be seen in figure. The model was established by K. E. BULLEN who, in turn, made use of the velocity-depth function of H. JEFFREYS. The most discussed part of the BULLEN model is the Byerly discontinuity of second order denoted at a depth of 410 km. In GUTENBERG's opinion the BYERLY discontinuity does not exist, instead of it a zone of low velocity beginning at a depth of 100-150 km, the low-velocity zone can be found in the upper mantle. This low-velocity zone plays an important role in the movements, which create the main features of the Earth, in the formation of mountains and in the development of the ridges of the oceans and of the rift valleys of deep seas, etc. From the decrease of velocity, one may presume here that the material has a lower rigidity and greater plasticity.

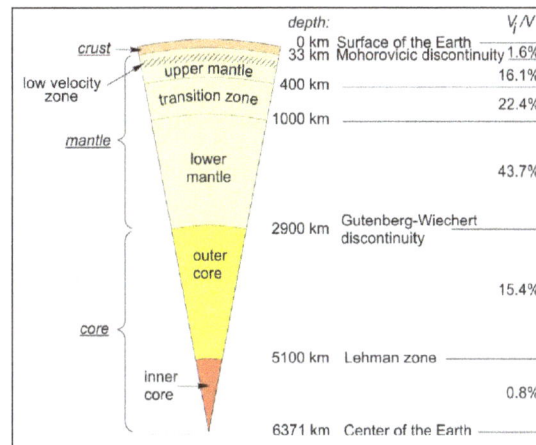

The layering within the earth.

The newest researches are directed to the recognition of the finer structure of the zones of the Earth. At the determination of the "fine structure" of the Earth, the observations of the free oscillations of the Earth and the underground nuclear explosions are utilized. These investigations also supported the assumption that a liquid-like zone is located around the centre of the Earth and a rigid but elastic mantle surrounds this zone. However, it was found at the same time that the theory of zonal structure is not completely correct because the material quality and the structure depend to a small extent also on the location of the point under the surface of the Earth. Essential differences exist, for example, between the deep structures below the oceans and below the continents. It has been proved by nuclear explosions that these differences are pre-sent even at depths of several hundred kilometers. Moreover, investigations of the free oscillations of the Earth have indicated that the core-mantle boundary is not flat but has a real "topography" in that the depths below the surface vary to a small extent from point to point.

Core, mantle, and crust are divisions based on composition. The crust makes up less than 1 percent of Earth by mass, consisting of oceanic crust and continental crust is often more felsic rock. The mantle is hot and represents about 68 percent of Earth's mass. Finally, the core is mostly iron metal. The core makes up about 31% of the Earth. Lithosphere and asthenosphere are divisions based on mechanical properties. The lithosphere is composed of both the crust and the portion of the upper mantle that behaves as a brittle, rigid solid. The asthenosphere is partially molten upper mantle material that behaves plastically and can flow. This animation by Earthquake shows the layers by composition and by mechanical properties.

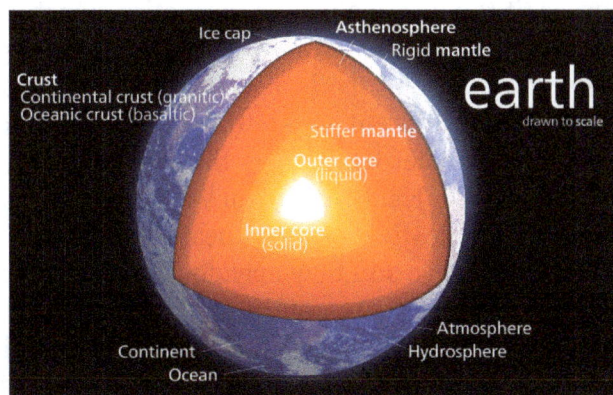

Crust and Lithosphere

Earth's outer surface is its crust; a cold, thin, brittle outer shell made of rock. The crust is very thin, relative to the radius of the planet. There are two very different types of crust, each with its own distinctive physical and chemical properties. Oceanic crust is composed of magma that erupts on the seafloor to create basalt lava flows or cools deeper down to create the intrusive igneous rock gabbro. Sediments, primarily muds and the shells of tiny sea creatures, coat the seafloor. Sediment is thickest near the shore where it comes off the continents in rivers and on wind currents. Continental crust is made up of many different types of igneous, metamorphic, and sedimentary rocks. The average composition is granite, which is much less dense than the mafic igneous rocks of the oceanic crust. Because it is thick and has relatively low density, continental crust rises higher on the mantle than oceanic crust, which sinks into the mantle to form basins. When filled with water, these basins form the planet's oceans. The lithosphere is the outermost mechanical layer, which behaves as a brittle, rigid solid. The lithosphere is about 100 kilometers thick. The definition of the lithosphere is based on how earth materials behave, so it includes the crust and the uppermost mantle, which are both brittle. Since it is rigid and brittle, when stresses act on the lithosphere, it breaks. This is what we experience as an earthquake.

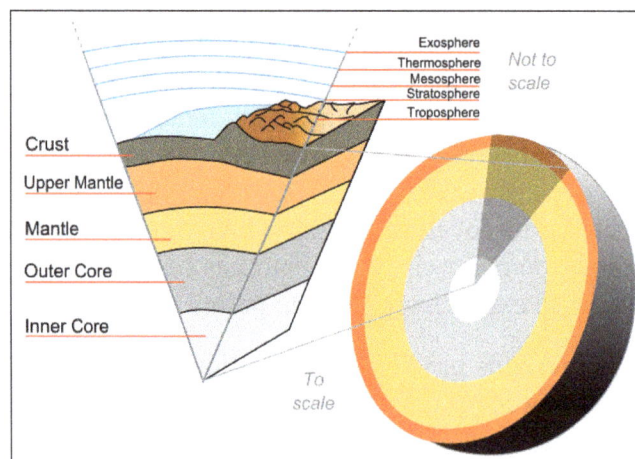

Mantle

The two most important things about the mantle are: (1) it is made of solid rock, and (2) it is hot. Scientists know that the mantle is made of rock based on evidence from seismic waves, heat flow, and meteorites. The properties fit the ultramafic rock peridotite, which is made of the iron- and magnesium-rich silicate minerals. Peridotite is rarely found at Earth's surface. Scientists know that the mantle is extremely hot because of the heat flowing outward from it and because of its physical properties. Heat flows in two different ways within the Earth: conduction and convection. Conduction is defined as the heat transfer that occurs through rapid collisions of atoms, which can only happen if the material is solid. Heat flows from warmer to cooler places until all are the same temperature. The mantle is hot mostly because of heat conducted from the core. Convection is the process of a material that can move and flow may develop convection currents. Convection in the mantle is the same as convection in a pot of water on a stove. Convection currents within Earth's mantle form as material near the core heats up. As the core heats the bottom layer of mantle material, particles move more rapidly, decreasing its density and causing it to rise. The rising material begins the convection current. When the warm material reaches the surface, it spreads horizontally. The material cools because it is no longer near the core. It eventually becomes cool and dense enough to sink back down into the mantle. At the bottom of the mantle, the material travels horizontally and is heated by the core. It reaches the location where warm mantle material rises, and the mantle convection cell is complete.

Convection in the mantle is the same as convection in a pot of water on a stove. Convection currents within Earth's mantle form as material near the core heats up. As the core heats the bottom layer of mantle material, particles move more rapidly, decreasing its density and causing it to rise. The rising material begins the convection current. When the warm material reaches the surface, it spreads horizontally. The material cools because it is no longer near the core. It eventually becomes cool and dense enough to sink back down into the mantle. At the bottom of the mantle, the material travels horizontally and is heated by the core. It reaches the location where warm mantle material rises, and the mantle convection cell is complete.

Core

At the planet's center lies a dense metallic core. Scientists know that the core is metal for a few reasons. The density of Earth's surface layers is much less than the overall density of the planet, as calculated from the planet's rotation. If the surface layers are less dense than average, then the interior must be denser than average. Calculations indicate that the core is about 85 percent iron metal with nickel metal making up much of the remaining 15 percent. Also, metallic meteorites are thought to be representative of the core. mIf Earth's core were not metal, the planet would not have a magnetic field. Metals such as iron are magnetic, but rock, which makes up the mantle and crust, is not. Scientists know that the outer core is liquid and the inner core is solid because S-waves stop at the inner core. The strong magnetic field is caused by convection in the liquid outer core. Convection currents in the outer core are due to heat from the even hotter inner core. The heat that keeps the outer core from solidifying is produced by the breakdown of radioactive elements in the inner core.

Crust

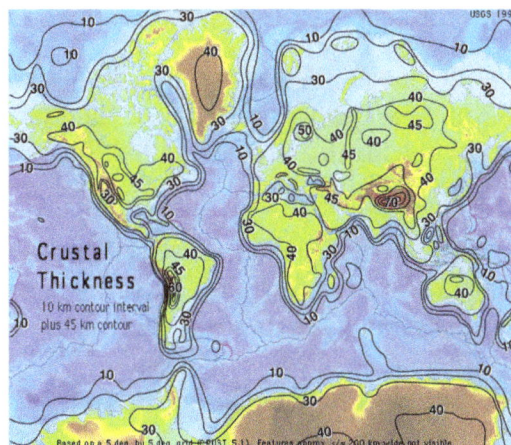

Earth crust

"Crust" describes the outermost shell of a terrestrial planet. Our planet's thin, 40-kilometer (25-mile) deep crust—just 1% of Earth's mass—contains all known life in the universe.

Earth has three layers: the crust, the mantle, and the core. The crust is made of solid rocks and minerals. Beneath the crust is the mantle, which is also mostly solid rocks and minerals, but punctuated by malleable areas of semi-solid magma. At the center of the Earth is a hot, dense metal core.

Earth's layers constantly interact with each other, and the crust and upper portion of the mantle are part of a single geologic unit called the lithosphere. The lithosphere's depth varies, and the Mohorovicic discontinuity (the Moho)—the boundary between the mantle and crust—does not exist at a uniform depth. Isostasy describes the physical, chemical, and mechanical differences between the mantle and crust that allow the crust to "float" on the more malleable mantle. Not all regions of Earth are balanced in isotactic equilibrium. Isotactic equilibrium depends on the density and thickness of the crust, and the dynamic forces at work in the mantle.

Just as the depth of the crust varies, so does its temperature. The upper crust withstands the ambient temperature of the atmosphere or ocean—hot in arid deserts and freezing in ocean trenches. Near the Moho, the temperature of the crust ranges from 200° Celsius (392° Fahrenheit) to 400° Celsius (752° Fahrenheit).

Crafting the Crust

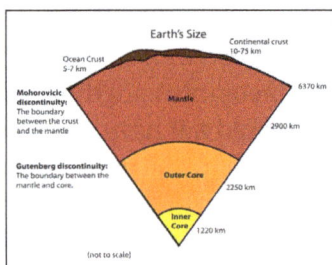

Earth size.

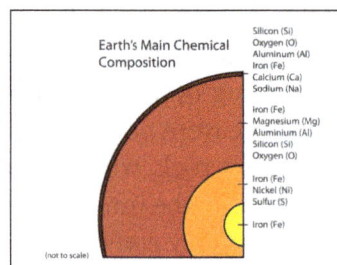

Earth chemical.

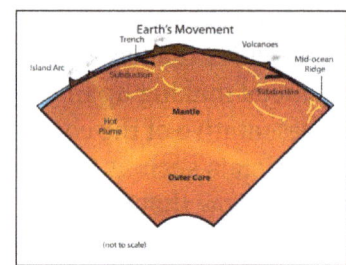

Earth movement.

Billions of years ago, the planetary blob that would become the Earth started out as a hot, viscous ball of rock. The heaviest material, mostly iron and nickel, sank to the center of the new planet and became its core. The molten material that surrounded the core was the early mantle.

Over millions of years, the mantle cooled. Water trapped inside minerals erupted with lava, a process called "outgassing." As more water was outgassed, the mantle solidified. Materials that initially stayed in their liquid phase during this process, called "incompatible elements," ultimately became Earth's brittle crust.

From mud and clay to diamonds and coal, Earth's crust is composed of igneous, metamorphic, and sedimentary rocks. The most abundant rocks in the crust are igneous, which are formed by the cooling of magma. Earth's crust is rich in igneous rocks such as granite and basalt. Metamorphic rocks have undergone drastic changes due to heat and pressure. Slate and marble are familiar metamorphic rocks. Sedimentary rocks are formed by the accumulation of material at Earth's surface. Sandstone and shale are sedimentary rocks.

Dynamic geologic forces created Earth's crust, and the crust continues to be shaped by the planet's movement and energy. Today, tectonic activity is responsible for the formation (and destruction) of crustal materials.

Earth's crust is divided into two types: oceanic crust and continental crust. The transition zone between these two types of crust is sometimes called the Conrad discontinuity. Silicates (mostly compounds made of silicon and oxygen) are the most abundant rocks and minerals in both oceanic and continental crust.

Oceanic Crust

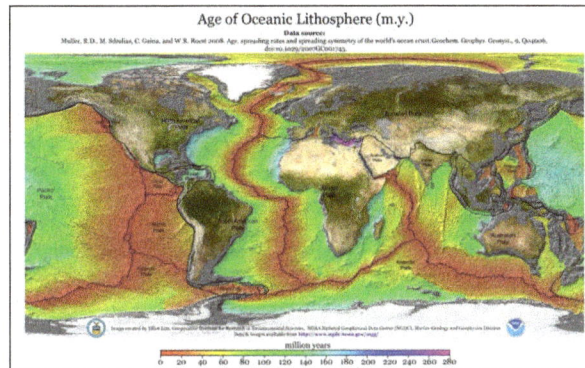

Oceanic crust.

 Oceanic crust, extending 5-10 kilometers (3-6 kilometers) beneath the ocean floor, is mostly composed of different types of basalts. Geologists often refer to the rocks of the oceanic crust as "sima." Sima stands for silicate and magnesium, the most abundant minerals in oceanic crust. (Basalts are a sima rocks.) Oceanic crust is dense, almost 3 grams per cubic centimeter (1.7 ounces per cubic inch).

Oceanic crust is constantly formed at mid-ocean ridges, where tectonic plates are tearing apart from each other. As magma that wells up from these rifts in Earth's surface cools, it becomes young oceanic crust. The age and density of oceanic crust increases with distance from mid-ocean ridges.

Just as oceanic crust is formed at mid-ocean ridges, it is destroyed in subduction zones. Subduction is the important geologic process in which a tectonic plate made of dense lithospheric material melts or falls below a plate made of less-dense lithosphere at a convergent plate boundary.

At convergent plate boundaries between continental and oceanic lithosphere, the dense oceanic lithosphere (including the crust) always subducts beneath the continental. In the northwestern United States, for example, the oceanic Juan de Fuca plate subducts beneath the continental North American plate. At convergent boundaries between two plates carrying oceanic lithosphere, the denser (usually the larger and deeper ocean basin) subducts. In the Japan Trench, the dense Pacific plate subducts beneath the less-dense Okhotsk plate.

As the lithosphere subducts, it sinks into the mantle, becoming more plastic and ductile. Through mantle convection, the rich minerals of the mantle may be ultimately "recycled" as they surface as crust-making lava at mid-ocean ridges and volcanoes.

Largely due to subduction, oceanic crust is much, much younger than continental crust. The oldest existing oceanic crust is in the Ionian Sea, part of the eastern Mediterranean basin. The seafloor of the Ionian Sea is about 270 million years old. (The oldest parts of continental crust, on the other hand, are more than 4 billion years old.)

Geologists collect samples of oceanic crust through drilling at the ocean floor, using submersibles, and studying ophiolites. Ophiolites are sections of oceanic crust that have been forced above sea level through tectonic activity, sometimes emerging as dikes in continental crust. Ophiolites are often more accessible to scientists than oceanic crust at the bottom of the ocean.

Continental Crust

Continental crust is mostly composed of different types of granites. Geologists often refer to the rocks of the continental crust as "sial." Sial stands for silicate and aluminum, the most abundant minerals in continental crust. Sial can be much thicker than sima (as thick as 70 kilometers kilometers (44 miles)), but also slightly less dense (about 2.7 grams per cubic centimeter (1.6 ounces per cubic inch)).

As with oceanic crust, continental crust is created by plate tectonics. At convergent plate boundaries, where tectonic plates crash into each other, continental crust is thrust up in the process of orogeny, or mountain-building. For this reason, the thickest parts of continental crust are at the world's tallest mountain ranges. Like icebergs, the tall peaks of the Himalayas and the Andes are only part of the region's continental crust—the crust extends unevenly below the Earth as well as soaring into the atmosphere.

Cratons are the oldest and most stable part of the continental lithosphere. These parts of the continental crust are usually found deep in the interior of most continents. Cratons are divided into two categories. Shields are cratons in which the ancient basement rock crops out into the atmosphere. Platforms are cratons in which the basement rock is buried beneath overlying sediment. Both shields and platforms provide crucial information to geologists about Earth's early history and formation.

Continental crust is almost always much older than oceanic crust. Because continental crust is rarely destroyed and recycled in the process of subduction, some sections of continental crust are nearly as old as the Earth itself.

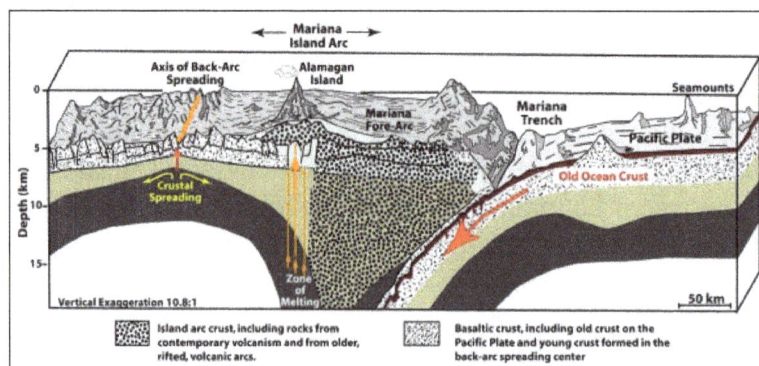

Mariana trench.

Mantle

The mantle is the mostly-solid bulk of Earth's interior. The mantle lies between Earth's dense, super-heated core and its thin outer layer, the crust. The mantle is about 2,900 kilometers (1,802 miles) thick, and makes up a whopping 84% of Earth's total volume.

As Earth began to take shape about 4.5 billion years ago, iron and nickel quickly separated from other rocks and minerals to form the core of the new planet. The molten material that surrounded the core was the early mantle.

Over millions of years, the mantle cooled. Water trapped inside minerals erupted with lava, a process called "outgassing." As more water was outgassed, the mantle solidified.

The rocks that make up Earth's mantle are mostly silicates—a wide variety of compounds that share a silicon and oxygen structure. Common silicates found in the mantle include olivine, garnet, and pyroxene. The other major type of rock found in the mantle is magnesium oxide. Other mantle elements include iron, aluminum, calcium, sodium, and potassium.

The temperature of the mantle varies greatly, from 1000 °Celsius (1832 °Fahrenheit) near its boundary with the crust, to 3700 °Celsius (6692 °Fahrenheit) near its boundary with the core. In the mantle, heat and pressure generally increase with depth. The geothermal gradient is a measurement of this increase. In most places, the geothermal gradient is about 25 °Celsius per kilometer of depth (1°Fahrenheit per 70 feet of depth).

The viscosity of the mantle also varies greatly. It is mostly solid rock, but less viscous at tectonic plate boundaries and mantle plumes. Mantle rocks there are soft and able to move plastically (over the course of millions of years) at great depth and pressure.

The transfer of heat and material in the mantle helps determine the landscape of Earth. Activity in the mantle drives plate tectonics, contributing to volcanoes, seafloor spreading, earthquakes, and orogeny (mountain-building).

The mantle is divided into several layers: the upper mantle, the transition zone, the lower mantle, and D" (D double-prime), the strange region where the mantle meets the outer core.

Upper Mantle

The upper mantle extends from the crust to a depth of about 410 kilometers (255 miles). The upper mantle is mostly solid, but its more malleable regions contribute to tectonic activity.

Two parts of the upper mantle are often recognized as distinct regions in Earth's interior: the lithosphere and the asthenosphere.

Lithosphere

The lithosphere is the solid, outer part of the Earth, extending to a depth of about 100 kilometers (62 miles). The lithosphere includes both the crust and the brittle upper portion of the mantle. The lithosphere is both the coolest and the most rigid of Earth's layers.

The most well-known feature associated with Earth's lithosphere is tectonic activity. Tectonic activity describes the interaction of the huge slabs of lithosphere called tectonic plates. The lithosphere is divided into 15 major tectonic plates: the North American, Caribbean, South American, Scotia, Antarctic, Eurasian, Arabian, African, Indian, Philippine, Australian, Pacific, Juan de Fuca, Cocos, and Nazca.

The division in the lithosphere between the crust and the mantle is called the Mohorovicic discontinuity, or simply the Moho. The Moho does not exist at a uniform depth, because not all regions of Earth are equally balanced in isostatic equilibrium. Isostasy describes the physical, chemical, and mechanical differences that allow the crust to "float" on the sometimes more malleable mantle. The Moho is found at about 8 kilometers (5 miles) beneath the ocean and about 32 kilometers (20 miles) beneath continents.

Different types of rocks distinguish lithospheric crust and mantle. Lithospheric crust is characterized by gneiss (continental crust) and gabbro (oceanic crust). Below the Moho, the mantle is characterized by peridotite, a rock mostly made up of the minerals olivine and pyroxene.

Asthenosphere

The asthenosphere is the denser, weaker layer beneath the lithospheric mantle. It lies between about 100 kilometers (62 miles) and 410 kilometers (255 miles) beneath Earth's surface. The temperature and pressure of the asthenosphere are so high that rocks soften and partly melt, becoming semi-molten.

The asthenosphere is much more ductile than either the lithosphere or lower mantle. Ductility measures a solid material's ability to deform or stretch under stress. The asthenosphere is generally more viscous than the lithosphere, and the lithosphere-asthenosphere boundary (LAB) is the point where geologists and rheologists—scientists who study the flow of matter—mark the difference in ductility between the two layers of the upper mantle.

The very slow motion of lithospheric plates "floating" on the asthenosphere is the cause of plate tectonics, a process associated with continental drift, earthquakes, the formation of mountains, and volcanoes. In fact, the lava that erupts from volcanic fissures is actually the asthenosphere itself, melted into magma.

Of course, tectonic plates are not really floating, because the asthenosphere is not liquid. Tectonic plates are only unstable at their boundaries and hot spots.

Transition Zone

From about 410 kilometers (255 miles) to 660 kilometers (410 miles) beneath Earth's surface, rocks undergo radical transformations. This is the mantle's transition zone.

In the transition zone, rocks do not melt or disintegrate. Instead, their crystalline structure changes in important ways. Rocks become much, much more dense.

The transition zone prevents large exchanges of material between the upper and lower mantle. Some geologists think that the increased density of rocks in the transition zone prevents subducted slabs from the lithosphere from falling further into the mantle. These huge pieces of tectonic plates stall in the transition zone for millions of years before mixing with other mantle rock and eventually returning to the upper mantle as part of the asthenosphere, erupting as lava, becoming part of the lithosphere, or emerging as new oceanic crust at sites of seafloor spreading.

Some geologists and ecologists, however, think subducted slabs can slip beneath the transition zone to the lower mantle. Other evidence suggests that the transition layer is permeable, and the upper and lower mantle exchange some amount of material.

Water

Perhaps the most important aspect of the mantle's transition zone is its abundance of water. Crystals in the transition zone hold as much water as all the oceans on Earth's surface.

Water in the transition zone is not "water" as we know it. It is not liquid, vapor, solid, or even plasma. Instead, water exists as hydroxide. Hydroxide is an ion of hydrogen and oxygen with a negative charge. In the transition zone, hydroxide ions are trapped in the crystalline structure of rocks such as ringwoodite and wadsleyite. These minerals are formed from olivine at very high temperatures and pressure.

Near the bottom of the transition zone, increasing temperature and pressure transform ringwoodite and wadsleyite. Their crystal structures are broken and hydroxide escapes as "melt." Melt particles flow upwards, toward minerals that can hold water. This allows the transition zone to maintain a consistent reservoir of water.

Geologists and rheologists think that water entered the mantle from Earth's surface during subduction. Subduction is the process in which a dense tectonic plate slips or melts beneath a more buoyant one. Most subduction happens as an oceanic plate slips beneath a less-dense plate. Along with the rocks and minerals of the lithosphere, tons of water and carbon are also transported to the mantle. Hydroxide and water are returned to the upper mantle, crust, and even atmosphere through mantle convection, volcanic eruptions, and seafloor spreading.

Lower Mantle

The lower mantle extends from about 660 kilometers (410 miles) to about 2,700 kilometers (1,678 miles) beneath Earth's surface. The lower mantle is hotter and denser than the upper mantle and transition zone.

The lower mantle is much less ductile than the upper mantle and transition zone. Although heat usually corresponds to softening rocks, intense pressure keeps the lower mantle solid.

Geologists do not agree about the structure of the lower mantle. Some geologists think that subducted slabs of lithosphere have settled there. Other geologists think that the lower mantle is entirely unmoving and does not even transfer heat by convection.

D Double-prime (D")

Beneath the lower mantle is a shallow region called D", or "d double-prime." In some areas, D" is a nearly razor-thin boundary with the outer core. In other areas, D" has thick accumulations of iron and silicates. In still other areas, geologists and seismologists have detected areas of huge melt.

The unpredictable movement of materials in D" is influenced by the lower mantle and outer core. The iron of the outer core influences the formation of a diaper, a dome-shaped geologic feature (igneous intrusion) where more fluid material is forced into brittle overlying rock. The iron diapir emits heat and may release a huge, bulging pulse of either material or energy—just like a Lava Lamp. This energy blooms upward, transferring heat to the lower mantle and transition zone, and maybe even erupting as a mantle plume.

At the base of the mantle, about 2,900 kilometers (1,802 miles) below the surface, is the core-mantle boundary, or CMB. This point, called the Gutenberg discontinuity, marks the end of the mantle and the beginning of Earth's liquid outer core.

Mantle Convection

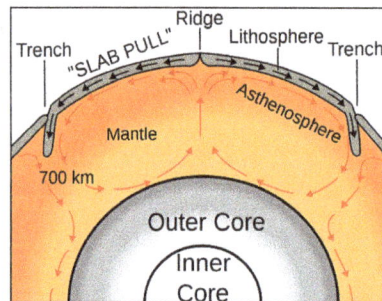

Mantle convection.

Mantle convection describes the movement of the mantle as it transfers heat from the white-hot core to the brittle lithosphere. The mantle is heated from below, cooled from above, and its overall temperature decreases over long periods of time. All these elements contribute to mantle convection.

Convection currents transfer hot, buoyant magma to the lithosphere at plate boundaries and hot spots. Convection currents also transfer denser, cooler material from the crust to Earth's interior through the process of subduction.

Earth's heat budget, which measures the flow of thermal energy from the core to the atmosphere, is dominated by mantle convection. Earth's heat budget drives most geologic processes on Earth, although its energy output is dwarfed by solar radiation at the surface.

Geologists debate whether mantle convection is "whole" or "layered." Whole-mantle convection describes a long, long recycling process involving the upper mantle, transition zone, lower mantle, and even D". In this model, the mantle convects in a single process. A subducted slab of lithosphere may slowly slip into the upper mantle and fall to the transition zone due to its relative density and coolness. Over millions of years, it may sink further into the lower mantle. Convection currents may then transport the hot, buoyant material in D" back through the other layers of the mantle. Some of that material may even emerge as lithosphere again, as it is spilled onto the crust through volcanic eruptions or seafloor spreading.

Layered-mantle convection describes two processes. Plumes of superheated mantle material may bubble up from the lower mantle and heat a region in the transition zone before falling back. Above the transition zone, convection may be influenced by heat transferred from the lower mantle as well as discrete convection currents in the upper mantle driven by subduction and seafloor spreading. Mantle plumes emanating from the upper mantle may gush up through the lithosphere as hot spots.

Mantle Plumes

A mantle plume is an upwelling of superheated rock from the mantle. Mantle plumes are the likely cause of "hot spots," volcanic regions not created by plate tectonics. As a mantle plume reaches the upper mantle, it melts into a diapir. This molten material heats the asthenosphere and lithosphere, triggering volcanic eruptions. These volcanic eruptions make a minor contribution to heat loss from Earth's interior, although tectonic activity at plate boundaries is the leading cause of such heat loss.

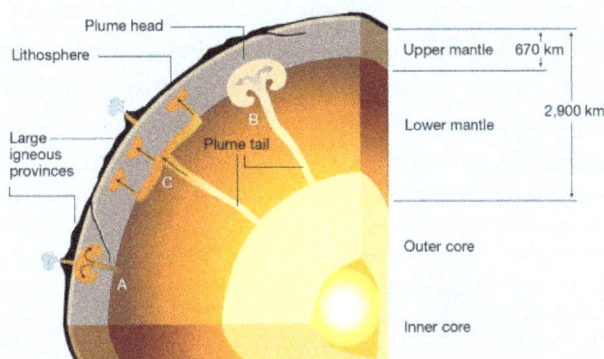

Mantle plumes

The Hawaiian hot spot, in the middle of the North Pacific Ocean, sits above a likely mantle plume. As the Pacific plate moves in a generally northwestern motion, the Hawaiian hot spot remains relatively fixed. Geologists think this has allowed the Hawaiian hot spot to create a series of volcanoes, from the 85-million-year-old Meiji Seamount near Russia's Kamchatka Peninsula, to the Loihi Seamount, a submarine volcano southeast of the "Big Island" of Hawaii. Loihi, a mere 400,000 years old, will eventually become the newest Hawaiian island.

Geologists have identified two so-called "superplumes." These superplumes, or large low shear velocity provinces (LLSVPs), have their origins in the melt material of D". The Pacific LLSVP influences geology throughout most of the southern Pacific Ocean (including the Hawaiian hot spot). The African LLSVP influences the geology throughout most of southern and western Africa.

Geologists think mantle plumes may be influenced by many different factors. Some may pulse, while others may be heated continually. Some may have a single diapir, while others may have multiple "stems." Some mantle plumes may arise in the middle of a tectonic plate, while others may be "captured" by seafloor spreading zones.

Some geologists have identified more than a thousand mantle plumes. Some geologists think mantle plumes don't exist at all. Until tools and technology allow geologists to more thoroughly explore the mantle, the debate will continue.

Exploring the Mantle

The mantle has never been directly explored. Even the most sophisticated drilling equipment has not reached beyond the crust.

Drilling all the way down to the Moho (the division between the Earth's crust and mantle) is an important scientific milestone, but despite decades of effort, nobody has yet succeeded. In

2005, scientists with the Integrated Ocean Drilling Project drilled 1,416 meters (4,644 feet) below the North Atlantic seafloor and claimed to have come within just 305 meters (1,000 feet) of the Moho.

Xenoliths

Many geologists study the mantle by analyzing xenoliths. Xenoliths are a type of intrusion—a rock trapped inside another rock.

The xenoliths that provide the most information about the mantle are diamonds. Diamonds form under very unique conditions: in the upper mantle, at least 150 kilometers (93 miles) beneath the surface. Above depth and pressure, the carbon crystallizes as graphite, not diamond. Diamonds are brought to the surface in explosive volcanic eruptions, forming "diamond pipes" of rocks called kimberlitic and lamprolites.

The diamonds themselves are of less interest to geologists than the xenoliths some contain. These intrusions are minerals from the mantle, trapped inside the rock-hard diamond. Diamond intrusions have allowed scientists to glimpse as far as 700 kilometers (435 miles) beneath Earth's surface—the lower mantle.

Xenolith studies have revealed that rocks in the deep mantle are most likely 3-billion-year old slabs of subducted seafloor. The diamond intrusions include water, ocean sediments, and even carbon.

Seismic Waves

Most mantle studies are conducted by measuring the spread of shock waves from earthquakes, called seismic waves. The seismic waves measured in mantle studies are called body waves, because these waves travel through the body of the Earth. The velocity of body waves differs with density, temperature, and type of rock.

There are two types of body waves: primary waves, or P-waves, and secondary waves, or S-waves. P-waves, also called pressure waves, are formed by compressions. Sound waves are P-waves—seismic P-waves are just far too low a frequency for people to hear. S-waves, also called shear waves, measure motion perpendicular to the energy transfer. S-waves are unable to transmit through fluids or gases.

Instruments placed around the world measure these waves as they arrive at different points on the Earth's surface after an earthquake. P-waves (primary waves) usually arrive first, while s-waves arrive soon after. Both body waves "reflect" off different types of rocks in different ways. This allows seismologists to identify different rocks present in Earth's crust and mantle far beneath the surface. Seismic reflections, for instance, are used to identify hidden oil deposits deep below the surface.

Sudden, predictable changes in the velocities of body waves are called "seismic discontinuities." The Moho is a discontinuity marking the boundary of the crust and upper mantle. The so-called "410-kilometer discontinuity" marks the boundary of the transition zone.

The Gutenberg discontinuity is more popularly known as the core-mantle boundary (CMB). At the CMB, S-waves, which can't continue in liquid, suddenly disappear, and P-waves are strongly refracted, or bent. This alerts seismologists that the solid and molten structure of the mantle has given way to the fiery liquid of the outer core.

Mantle Maps

Cutting-edge technology has allowed modern geologists and seismologists to produce mantle maps. Most mantle maps display seismic velocities, revealing patterns deep below Earth's surface.

Geoscientists hope that sophisticated mantle maps can plot the body waves of as many as 6,000 earthquakes with magnitudes of at least 5.5. These mantle maps may be able to identify ancient slabs of subducted material and the precise position and movement of tectonic plates. Many geologists think mantle maps may even provide evidence for mantle plumes and their structure.

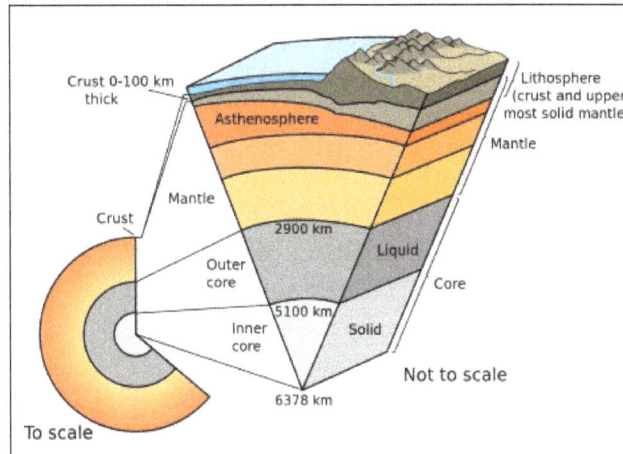

Cutaway earth

Core

Earth's core is the very hot, very dense center of our planet. The ball-shaped core lies beneath the cool, brittle crust and the mostly-solid mantle. The core is found about 2,900 kilometers (1,802 miles) below Earth's surface, and has a radius of about 3,485 kilometers (2,165 miles).

Planet Earth is older than the core. When Earth was formed about 4.5 billion years ago, it was a uniform ball of hot rock. Radioactive decay and leftover heat from planetary formation (the collision, accretion, and compression of space rocks) caused the ball to get even hotter. Eventually, after about 500 million years, our young planet's temperature heated to the melting point of iron—about 1,538° Celsius (2,800° Fahrenheit). This pivotal moment in Earth's history is called the iron catastrophe.

The iron catastrophe allowed greater, more rapid movement of Earth's molten, rocky material. Relatively buoyant material, such as silicates, water, and even air, stayed close to the planet's exterior. These materials became the early mantle and crust. Droplets of iron, nickel, and other heavy metals gravitated to the center of Earth, becoming the early core. This important process is called planetary differentiation.

Earth's core is the furnace of the geothermal gradient. The geothermal gradient measures the increase of heat and pressure in Earth's interior. The geothermal gradient is about 25 °Celsius per kilometer of depth (1° Fahrenheit per 70 feet). The primary contributors to heat in the core are the decay of radioactive elements, leftover heat from planetary formation, and heat released as the liquid outer core solidifies near its boundary with the inner core.

Unlike the mineral-rich crust and mantle, the core is made almost entirely of metal—specifically, irons and nickel. The shorthand used for the core's iron-nickel alloys is simply the elements' chemical symbols—NiFe.

Elements that dissolve in iron, called siderophiles, are also found in the core. Because these elements are found much more rarely on Earth's crust, many siderophiles are classified as "precious metals." Siderophile elements include gold, platinum, and cobalt.

Another key element in Earth's core is sulfur—in fact 90% of the sulfur on Earth is found in the core. The confirmed discovery of such vast amounts of sulfur helped explain a geologic mystery: If the core was primarily NiFe, why wasn't it heavier? Geoscientists speculated that lighter elements such as oxygen or silicon might have been present. The abundance of sulfur, another relatively light element, explained the conundrum.

Although we know that the core is the hottest part of our planet, its precise temperatures are difficult to determine. The fluctuating temperatures in the core depend on pressure, the rotation of the Earth, and the varying composition of core elements. In general, temperatures range from about 4,400 °Celsius (7,952 °Fahrenheit) to about 6,000 °Celsius (10,800 °Fahrenheit).

The core is made of two layers: the outer core, which borders the mantle, and the inner core. The boundary separating these regions is called the Bullen discontinuity.

Outer Core

The outer core, about 2,200 kilometers (1,367 miles) thick, is mostly composed of liquid iron and nickel. The NiFe alloy of the outer core is very hot, between 4,500° and 5,500 °Celsius (8,132° and 9,932 °Fahrenheit).

The liquid metal of the outer core has very low viscosity, meaning it is easily deformed and malleable. It is the site of violent convection. The churning metal of the outer core creates and sustains Earth's magnetic field.

The hottest part of the core is actually the Bullen discontinuity, where temperatures reach 6,000° Celsius (10,800 °Fahrenheit)—as hot as the surface of the sun.

Inner Core

The inner core is a hot, dense ball of (mostly) iron. It has a radius of about 1,220 kilometers (758 miles). Temperature in the inner core is about 5,200 °Celsius (9,392 °Fahrenheit). The pressure is nearly 3.6 million atmosphere (atm).

The temperature of the inner core is far above the melting point of iron. However, unlike the outer core, the inner core is not liquid or even molten. The inner core's intense pressure—the entire rest of the planet and its atmosphere—prevents the iron from melting. The pressure and density are simply too great for the iron atoms to move into a liquid state. Because of this unusual set of circumstances, some geophysicists prefer to interpret the inner core not as a solid, but as a plasma behaving as a solid.

The liquid outer core separates the inner core from the rest of the Earth, and as a result, the inner core rotates a little differently than the rest of the planet. It rotates eastward, like the surface, but it's a little faster, making an extra rotation about every 1,000 years.

Geoscientists think that the iron crystals in the inner core are arranged in an "hcp" (hexagonal close-packed) pattern. The crystals align north-south, along with Earth's axis of rotation and magnetic field.

The orientation of the crystal structure means that seismic waves—the most reliable way to study the core—travel faster when going north-south than when going east-west. Seismic waves travel four seconds faster pole-to-pole than through the Equator.

Growth in the Inner Core

As the entire Earth slowly cools, the inner core grows by about a millimeter every year. The inner core grows as bits of the liquid outer core solidify or crystallize. Another word for this is "freezing," although it's important to remember that iron's freezing point more than 1,000 °Celsius (1,832 °Fahrenheit).

The growth of the inner core is not uniform. It occurs in lumps and bunches, and is influenced by activity in the mantle.

Growth is more concentrated around subduction zones—regions where tectonic plates are slipping from the lithosphere into the mantle, thousands of kilometers above the core. Subducted plates draw heat from the core and cool the surrounding area, causing increased instances of solidification.

Growth is less concentrated around "superplumes" or LLSVPs. These ballooning masses of superheated mantle rock likely influence "hot spot" volcanism in the lithosphere, and contribute to a more liquid outer core.

The core will never "freeze over." The crystallization process is very slow, and the constant radioactive decay of Earth's interior slows it even further. Scientists estimate it would take about 91 billion years for the core to completely solidify—but the sun will burn out in a fraction of that time (about 5 billion years).

Core Hemispheres

Just like the lithosphere, the inner core is divided into eastern and western hemispheres. These hemispheres don't melt evenly, and have distinct crystalline structures.

The western hemisphere seems to be crystallizing more quickly than the eastern hemisphere. In fact, the eastern hemisphere of the inner core may actually be melting.

Inner Core

Geoscientists recently discovered that the inner core itself has a core—the inner inner core. This strange feature differs from the inner core in much the same way the inner core differs from the outer core. Scientists think that a radical geologic change about 500 million years ago caused this inner inner core to develop.

The crystals of the inner inner core are oriented east-west instead of north-south. This orientation is not aligned with either Earth's rotational axis or magnetic field. Scientists think the iron crystals may even have a completely different structure (not hcp), or exist at a different phase.

Magnetism

Earth's magnetic field is created in the swirling outer core. Magnetism in the outer core is about 50 times stronger than it is on the surface.

It might be easy to think that Earth's magnetism is caused by the big ball of solid iron in the middle. But in the inner core, the temperature is so high the magnetism of iron is altered. Once this temperature, called the Curie point, is reached, the atoms of a substance can no longer align to a magnetic point.

Dynamo Theory

Some geoscientists describe the outer core as Earth's "geodynamo." For a planet to have a geodynamo, it must rotate, it must have a fluid medium in its interior, the fluid must be able to conduct electricity, and it must have an internal energy supply that drives convection in the liquid.

Variations in rotation, conductivity, and heat impact the magnetic field of a geodynamo. Mars, for instance, has a totally solid core and a weak magnetic field. Venus has a liquid core, but rotates too slowly to churn significant convection currents. It, too, has a weak magnetic field. Jupiter, on the other hand, has a liquid core that is constantly swirling due to the planet's rapid rotation.

Earth is the "Goldilocks" geodynamo. It rotates steadily, at a brisk 1,675 kilometers per hour (1,040 miles per hour) at the Equator. Coriolis forces, an artifact of Earth's rotation, cause convection currents to be spiral. The liquid iron in the outer core is an excellent electrical conductor, and creates the electrical currents that drive the magnetic field.

The energy supply that drives convection in the outer core is provided as droplets of liquid iron freeze onto the solid inner core. Solidification releases heat energy. This heat, in turn, makes the remaining liquid iron more buoyant. Warmer liquids spiral upward, while cooler solids spiral downward under intense pressure: convection.

Earth's Magnetic Field

Earth's magnetic field is crucial to life on our planet. It protects the planet from the charged particles of the solar wind. Without the shield of the magnetic field, the solar wind would strip Earth's atmosphere of the ozone layer that protects life from harmful ultraviolet radiation.

Although Earth's magnetic field is generally stable, it fluctuates constantly. As the liquid outer core moves, for instance, it can change the location of the magnetic North and South Poles. The magnetic North Pole moves up to 64 kilometers (40 miles) every year.

Fluctuations in the core can cause Earth's magnetic field to change even more dramatically. Geomagnetic pole reversals, for instance, happen about every 200,000 to 300,000 years. Geomagnetic pole reversals are just what they sound like: a change in the planet's magnetic poles, so that the magnetic North and South Poles are reversed. These "pole flips" are not catastrophic—scientists have noted no real changes in plant or animal life, glacial activity, or volcanic eruptions during previous geomagnetic pole reversals.

Studying the Core

Geoscientists cannot study the core directly. All information about the core has come from sophisticated reading of seismic data, analysis of meteorites, lab experiments with temperature and pressure, and computer modeling.

Most core research has been conducted by measuring seismic waves, the shock waves released by earthquakes at or near the surface. The velocity and frequency of seismic body waves changes with pressure, temperature, and rock composition.

In fact, seismic waves helped geoscientists identify the structure of the core itself. In the late 19th century, scientists noted a "shadow zone" deep in the Earth, where a type of body wave called an s-wave either stopped entirely or was altered. S-waves are unable to transmit through fluids or gases. The sudden "shadow" where s-waves disappeared indicated that Earth had a liquid layer.

In the 20th century, geoscientists discovered an increase in the velocity of p-waves, another type of body wave, at about 5,150 kilometers (3,200 miles) below the surface. The increase in velocity corresponded to a change from a liquid or molten medium to a solid. This proved the existence of a solid inner core.

Meteorites, space rocks that crash to Earth, also provide clues about Earth's core. Most meteorites are fragments of asteroids, rocky bodies that orbit the sun between Mars and Jupiter. Asteroids formed about the same time, and from about the same material, as Earth. By studying iron-rich chondrite meteorites, geoscientists can get a peek into the early formation of our solar system and Earth's early core.

In the lab, the most valuable tool for studying forces and reactions at the core is the diamond anvil cell. Diamond anvil cells use the hardest substance on Earth (diamonds) to simulate the incredibly high pressure at the core. The device uses an x-ray laser to simulate the core's temperature. The laser is beamed through two diamonds squeezing a sample between them.

Complex computer modeling has also allowed scientists to study the core. In the 1990s, for instance, modeling beautifully illustrated the geodynamo—complete with pole flips.

Continents and Oceans

Continental Drift Theory

Continental drift was a revolutionary scientific theory developed in the years 1908-1912 by Alfred Wegener, a German meteorologist, climatologist, and geophysicist, that put forth the hypothesis that the continents had all originally been a part of one enormous landmass or supercontinent about 240 million years ago before breaking apart and drifting to their current locations. Based on the work of previous scientists who had theorized about horizontal movement of the continents over the Earth's surface during different periods of geologic time, and based on his own observations drawing from different fields of science, Wegener postulated that about 200 million years ago, a supercontinent that he called Pangaea (which means "all lands" in Greek) began to break up. Over millions of years the pieces separated, first into two smaller supercontinents, Laurasia and Gondwanaland, during the Jurassic period and then by the end of the Cretaceous period into the continents we know today.

Wegener first presented his ideas in 1912 and then published them in 1915 in his controversial book, "The Origins of Continents and Oceans," which was received with great skepticism and even hostility. He revised and published subsequent editions of his book in 1920, 1922, and 1929. The book (Dover translation of the 1929 fourth German edition) is still available today on Amazon and elsewhere.

Wegener's theory, although not completely correct, and by his own admission, incomplete, sought to explain why similar species of animals and plants, fossil remains, and rock formations exist on disparate lands separated by great distances of sea. It was an important and influential step that ultimately led to the development of the theory of plate tectonics, which is how scientists understand the structure, history, and dynamics of the Earth's crust.

Opposition to Continental Drift Theory

There was much opposition to Wegener's theory for several reasons. For one, he was not an expert in the field of science in which he was making a hypothesis, and for another, his radical theory threatened conventional and accepted ideas of the time. Furthermore, because he was making observations that were multidisciplinary, there were more scientists to find fault with them.

There were also alternative theories to counter Wegener's continental drift theory. A commonly held theory to explain the presence of fossils on disparate lands was that there was once a network of land bridges connecting the continents that had sunk into the sea as part of a general cooling and contraction of the earth. Wegener, however, refuted this theory maintaining that continents were made of a less dense rock than that of the deep-sea floor and so would have risen to the surface again once the force weighing them down had been lifted. Since this had not occurred, according to Wegener, the only logical alternative was that the continents themselves had been joined and had since drifted apart.

Another theory was that the fossils of temperate species found in the arctic regions were carried there by warm water currents. Scientists debunked these theories, but at the time they helped stall Wegener's theory from gaining acceptance.

In addition, many of the geologists who were Wegener's contemporaries were constructionists. They believed that the Earth was in the process of cooling and shrinking, an idea they used to explain the formation of mountains, much like wrinkles on a prune. Wegener, though, pointed out that if this were true, mountains would be scattered evenly all over the Earth's surface rather than lined up in narrow bands, usually at the edge of a continent. He also offered a more plausible explanation for mountain ranges. He said they formed when the edge of a drifting continent crumpled and folded — as when India hit Asia and formed the Himalayas.

One of the biggest flaws of Wegener's continental drift theory was that he did not have a viable explanation for how continental drift could have occurred. He proposed two different mechanisms, but each was weak and could be disproven. One was based on the centrifugal force caused by the rotation of the Earth, and the other was based on the tidal attraction of the sun and the moon.

Though much of what Wegener theorized was correct, the few things that were wrong were held against him and prevented him from seeing his theory accepted by the scientific community during his lifetime. However, what he got right paved the way for plate tectonics theory.

Evidence

Evidence in Support of Continental Drift

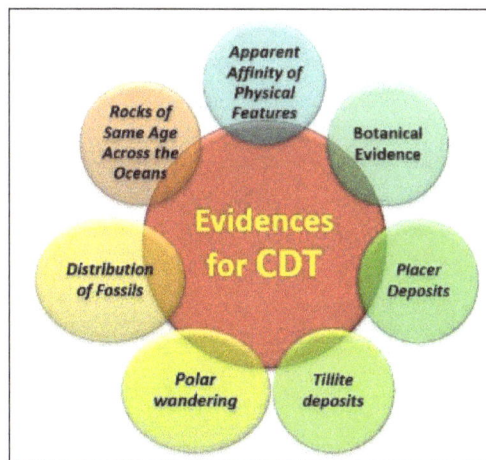

Apparent Affinity of Physical Features

- South America and Africa seem to fit in with each other, especially, the bulge of Brazil fits into the Gulf of Guinea.

- Greenland seems to fit in well with Ellesmere and Baffin islands.

- The west coast of India, Madagascar and Africa seem to have been joined.

- North and South America on one side and Africa and Europe on the other fit along the mid-Atlantic ridge.

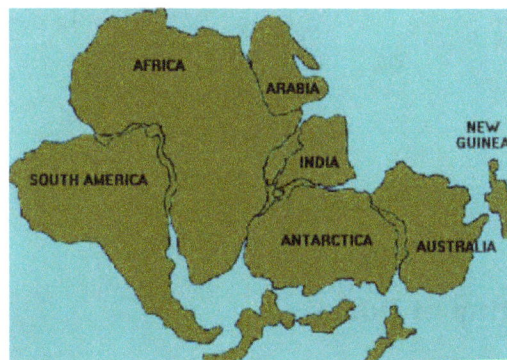

- The Caledonian and Hercynian mountains of Europe and the Appalachians of USA seem to be one continuous series.

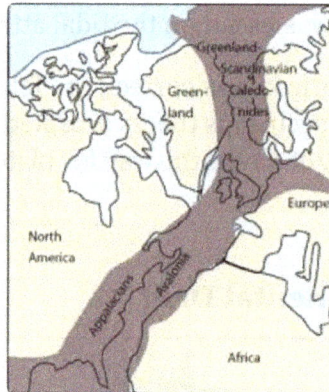
Caledonian mountains.

Caledonian and Hercynian mountains.

Criticism:

- Coastlines are a temporary feature and are liable to change.

- Several other combinations of fitting in of landforms could be attempted.

- Continental Drift Theory shifts India's position too much to the south, distorting its relation with the Mediterranean Sea and the Alps.

- The mountains do not always exhibit geological affinity.

Causes of Drift

Gravity of the earth, buoyancy of the seas and the tidal currents were given as the main factors causing the drift, by Wegener.

Criticism:

This is illogical because for these factors to be able to cause a drift of such a magnitude, they will have to be millions of times stronger.

Polar Wandering (Shifting of Poles)

- The poles drifted constantly.

Criticism:

- Poles may have shifted, not necessarily the continents (don't think deep).

Distribution of Fossils

- The observations that Lemurs occur in India, Madagascar and Africa led some to consider a contiguous landmass "Lemuria" linking these three landmasses.

- Mesosaurus was a small reptile adapted to shallow brackish water. The skeletons of these are found only in South Africa and Iraver formations of Brazil. The two localities presently are 4,800 km apart with an ocean in between them.

Botanical Evidence

- Presence of glossopteris vegetation in carboniferous rocks of India, Australia, South Africa, Falkland Islands (Overseas territory of UK), Antarctica, etc. can be explained on the basis of the fact that parts were linked in the past.

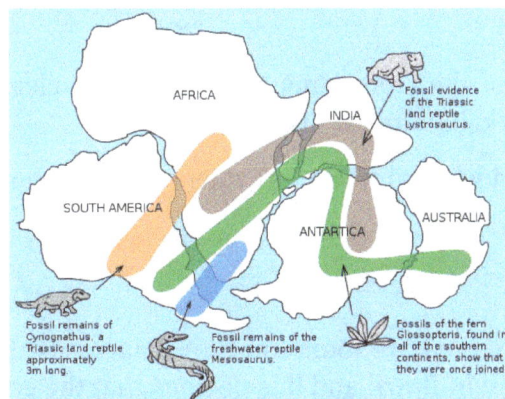

Criticism:

- Such vegetation is also found in the northern parts like Afghanistan, Iran and Siberia.

- Similar vegetation found in unrelated parts of the world.

Rocks of Same Age across the Oceans

- The belt of ancient rocks of 2,000 million years from Brazil coast matches with those from western Africa.

Criticism:

- Rocks of same age and similar characteristics are found in other parts of the world too.

Tillite Deposits

- It is the sedimentary rock formed out of deposits of glaciers. The Gondwana system of sediments from India is known to have its counter parts in six different landmasses of the Southern Hemisphere.

- At the base the system has thick Tillite indicating extensive and prolonged glaciation. Counter parts of this succession are found in Africa, Falkland Island, Madagascar, Antarctica and Australia besides India.

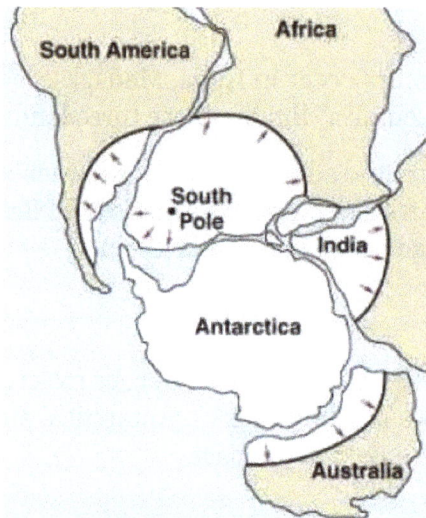

- Overall resemblance of the Gondwana type sediments clearly demonstrates that these landmasses had remarkably similar histories.

- The glacial Tillite provides unambiguous evidence of palaeoclimates and also of drifting of continents.

Placer Deposits

- Rich placer deposits of gold are found on the Ghana coast (West Africa) but the source (gold bearing veins) are in Brazil and it is obvious that the gold deposits of the Ghana are derived from the Brazil plateau when the two continents lay side by side.

Ocean Floor Configuration

The ocean basins are the result of tectonic forces and processes. All of the ocean basins were formed from volcanic rock that was released from fissures located at the mid-oceanic ridges. The oldest rocks found in these basins are approximately 200 million years old. This is a lot younger than the oldest continental rocks which have ages greater than 4 billion years. The reason for this discrepancy is simple. Tectonic processes destroy old oceanic rocks! Oceanic rock is returned to the Earth's mantle when oceanic crust is subducted. Many of these subduction zones occur at the continental margins where oceanic crust meets continental crust. Subduction also creates the ocean's deep trenches.

Topography of the Ocean Basins

The ocean basins are not featureless Earth surfaces. Much of our knowledge about the topographic features that exist here are derived from the following technologies: seismic surveying; echo sounder; side-scan sonar; and the measurement of the height of sea surfaces by satellites. Most of the general information concerning the depth of the ocean basins were made after World War

I when the echo sounder was developed for military purposes. This instrument accurately determines the time between the emission of a strong acoustic pulse and the detection of its echo. Using this principle scientists can determine the distance from the sounder to the ocean bottom.

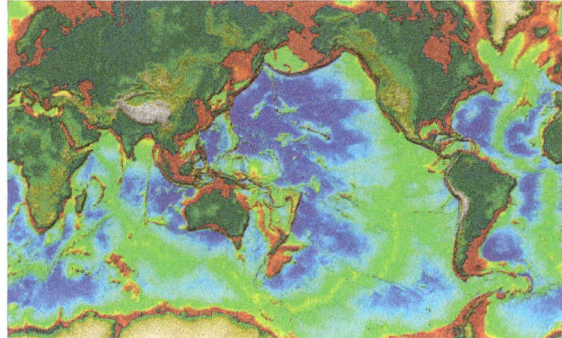

The above image displays the topography of the Earth's terrestrial land surface and ocean basins. Data for the image comes from satellite altimetry and ship depth soundings, and U.S. Geological Survey digital elevation maps (DEM) of the Earth's land surface. In the ocean basin, the gradation from red to yellow to green to blue indicates increasing depth. A number of topographic features associated with the ocean basin can be seen in this image. The red area that borders the various landmasses is the continental shelf. This feature is structurally part of the continental landmasses despite the fact that it is under water. The yellow to green zone around the continental shelf is the continental slope and continental rise. The blue region in the various ocean basins constitutes the ocean floor. In the center of ocean basins, the mid-oceanic ridges can be seen with a color ranging from green to yellow to orange.

Some of the dominant topographic features associated with the ocean basins include:

Continental shelf is a shallow (average depth 130 meters) gently sloping part of the continental crust that borders the continents. The extent of this feature varies from tens of meters to a maximum width of about 1300 kilometers.

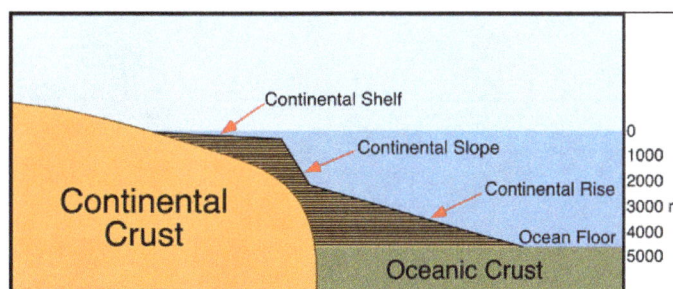

Marginal features found at the interface of the continents and the ocean basins.

The continental slope extends from the continental shelf at an average depth of about 135 meters. The base of this steeply sloping (from 1 to 25°, average about 4°) topographic feature occurs at a depth of approximately 2000 meters, marking the edge of the continents. The width of the slope varies from 20 to 100 kilometers. Both the continental shelf and slope are considered structurally part of the continents, even though they are below the sea surface. The boundary between the continental slope and shelf is called the continental shelf break.

Submarine canyons are V-shaped canyons cut into the continental slope to a depth of up to 1200 meters. The submarine canyons are cut perpendicular to the running direction of the continental slope. Many canyons are associated with major rivers such as the Congo, Hudson, and others.

The continental rise is found at the base of the continental slope. The depth of the rise ranges from 2000 to 5000 meters deep. Its breadth is up 300 kilometers wide. This feature was created by the merging of accumulated deposits at the mouths of the many submarine canyons. Each canyon's thick fan-shaped sedimentary deposit is called an abyssal fans.

The ocean floor is found at the base of the continental rise in water 4000 to 6000 meters. The ocean floor accounts for nearly 30% of the Earth's surface. The composition of the ocean floor consists of a relatively thin layer (on average 5 kilometers thick) of basaltic rock with an average density of 3.0 grams per cubic centimeter (continents - granite rocks - density 2.7 grams per cubic centimeter).

Numerous volcanoes populate the floor of the ocean basins. Scientists estimate that there are approximately 10,000 volcanoes on the ocean floor.

Mid-oceanic ridge is normally found rising above the ocean floor at the center of the ocean basins. These features are involved in the generation of new oceanic crust from volcanic fissures produced by mantle up-welling. Some volcanic islands are part of the mid-ocean ridge system (Iceland). The mid-oceanic ridge constitutes 23% of the Earth's surface. In the center of the mid-oceanic ridge is a rift valley, between 30 to 50 kilometers wide, that dissects 1000 to 3000 meters deep into the ridge system.

Ocean trenches are long, narrow, steep-sided depressions found on the ocean floor that contain the greatest depths in the ocean (11,000 meters - western Pacific). There are 26 oceanic trenches in the world: 3 in the Atlantic Ocean, 1 in the Indian Ocean, and 22 in the Pacific Ocean. Generally, the trenches mark the transition between continents and ocean basins, especially in the Pacific basin. Trenches are also the tectonic areas.

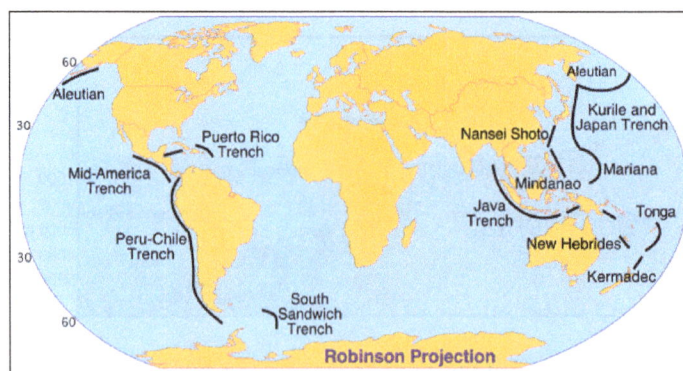

Major ocean trenches of the world. The Mariana Trench is the deepest at 11,020 meters below sea-level.

Ocean Basin Configuration

The current spatial configuration of the ocean basins is a by product of plate tectonics. The creation of new oceanic crust at the mid-oceanic ridge moves the continents across the Earth's surface and creates zones of subduction. At the areas of subduction, oceanic crust is forced into the mantle after it collides with continental crust. Over the past 200 million years, the Atlantic basin has been the most active area of oceanic crust creation. The Atlantic ocean formed about 200 million years ago as the Pangaean continent began rifting apart. 180 million years ago, North American separated from South America and Africa. North America then joined with Eurasia creating Laurasia. By 135 million years ago, South America began separating from Africa. North America and Eurasia split a few million years after.

Sea Floor Spreading

Seafloor spreading, theory that oceanic crust forms along submarine mountain zones, known collectively as the mid-ocean ridge system, and spreads out laterally away from them. This idea played a pivotal role in the development of plate tectonics, a theory that revolutionized geologic thought during the last quarter of the 20th century.

Shortly after the conclusion of World War II, sonar-equipped vessels crisscrossed the oceans collecting ocean-depth profiles of the seafloor beneath them. The survey data was used to create three-dimensional relief maps of the ocean floor, and, by 1953, American oceanic cartographer Marie Tharp had created the first of several maps that revealed the presence of an underwater mountain range more than 16,000 km (10,000 miles) long in the Atlantic—the Mid-Atlantic Ridge.

The seafloor spreading hypothesis was proposed by the American geophysicist Harry H. Hess in 1960. On the basis of Tharp's efforts and other new discoveries about the deep-ocean floor, Hess postulated that molten material from Earth's mantle continuously wells up along the crests of the mid-ocean ridges that wind for nearly 80,000 km (50,000 miles) through all the world's oceans. As the magma cools, it is pushed away from the flanks of the ridges. This spreading creates a successively younger ocean floor, and the flow of material is thought to bring about the migration, or drifting apart, of the continents. The continents bordering the Atlantic Ocean, for example, are believed to be moving away from the Mid-Atlantic Ridge at a rate of 1–2 cm (0.4–0.8 inch) per year, thus increasing the breadth of the ocean basin by twice that amount. Wherever continents are bordered by deep-sea trench systems, as in the Pacific Ocean, the ocean floor is plunged downward, under thrusting the continents and ultimately reentering and dissolving in Earth's mantle, from which it had originated.

A veritable legion of evidence supports the seafloor spreading hypothesis. Studies conducted with thermal probes, for example, indicate that the heat flow through bottom sediments is generally comparable to that through the continents except over the mid-ocean ridges, where at some sites the heat flow measures three to four times the normal value. The anomalously high values are considered to reflect the intrusion of molten material near the crests of the ridges. Research has also revealed that the ridge crests are characterized by anomalously low seismic wave velocities, which can be attributed to thermal expansion and micro fracturing associated with the upwelling magma.

Investigations of oceanic magnetic anomalies have further corroborated the seafloor spreading hypothesis. Such studies have shown that the strength of the geomagnetic field is alternately anomalously high and low with increasing distance away from the axis of the mid-ocean ridge system. The anomalous features are nearly symmetrically arranged on both sides of the axis and parallel the axis, creating bands of parallel anomalies.

Seafloor spreading in three ocean basins.

Measurements of the thickness of marine sediments and absolute age determinations of such bottom material have provided additional evidence for seafloor spreading. The oldest sediments so far recovered by a variety of methods—including coring, dredging, and deep-sea drilling—date only to the Jurassic Period, not exceeding about 200 million years in age. Such findings are incompatible with the doctrine of the permanency of the ocean basins that had prevailed among Earth scientists for so many years.

Plate Tectonics

Developing the Theory

In line with other previous and contemporaneous proposals, in 1912 the meteorologist Alfred Wegener amply described what he called continental drift, and the scientific debate started that would end up fifty years later in the theory of plate tectonics. Starting from the idea (also expressed by his forerunners) that the present continents once formed a single land mass (which was called Pangea later on) that drifted apart, thus releasing the continents from the Earth's mantle and likening them to "icebergs" of low density granite floating on a sea of denser basalt.

Supporting evidence for the idea came from the dove-tailing outlines of South America's east coast and Africa's west coast, and from the matching of the rock formations along these edges. Confirmation of their previous contiguous nature also came from the fossil plants Glossopteris and Gangamopteris, and the therapsid or mammal-like reptile Lystrosaurus, all widely distributed over South America, Africa, Antarctica, India and Australia. The evidence for such an erstwhile joining of these continents was patent to field geologists working in the southern hemisphere. The South African Alex du Toit put together a mass of such information in his 1937 publication, and went further than Wegener in recognising the strong links between the Gondwana fragments.

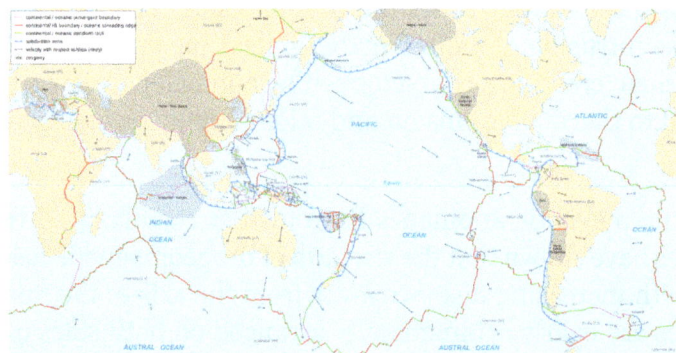

Detailed map showing the tectonic plates with their movement vectors.

But without detailed evidence and a force sufficient to drive the movement, the theory was not generally accepted: the Earth might have a solid crust and mantle and a liquid core, but there seemed to be no way that portions of the crust could move around. Distinguished scientists, such as Harold Jeffreys and Charles Schuchert, were outspoken critics of continental drift.

Despite much opposition, the view of continental drift gained support and a lively debate started between "drifters" or "mobilists" (proponents of the theory) and "fixists" (opponents). During the 1920s, 1930s and 1940s, the former reached important milestones proposing that convection currents might have driven the plate movements, and that spreading may have occurred below the sea within the oceanic crust. Concepts close to the elements now incorporated in plate tectonics were proposed by geophysicists and geologists (both fixists and mobilists) like Vening-Meinesz, Holmes, and Umbgrove.

One of the first pieces of geophysical evidence that was used to support the movement of lithospheric plates came from paleomagnetism. This is based on the fact that rocks of different ages show a variable magnetic field direction, evidenced by studies since the mid–nineteenth century. The magnetic north and south poles reverse through time, and, especially important in paleotectonic studies, the relative position of the magnetic north pole varies through time. Initially, during the first half of the twentieth century, the latter phenomenon was explained by introducing what was called "polar wander", i.e., it was assumed that the north pole location had been shifting through time. An alternative explanation, though, was that the continents had moved (shifted and rotated) relative to the north pole, and each continent, in fact, shows its own "polar wander path". During the late 1950s it was successfully shown on two occasions that these data could show the validity of continental drift.

The second piece of evidence in support of continental drift came during the late 1950s and early 60s from data on the bathymetry of the deep ocean floors and the nature of the oceanic crust such as magnetic properties and, more generally, with the development of marine geology which gave evidence for the association of seafloor spreading along the mid-oceanic ridges and magnetic field reversals, published between 1959 and 1963 by Heezen, Dietz, Hess, Mason, Vine & Matthews, and Morley.

Simultaneous advances in early seismic imaging techniques in and around Wadati-Beni off zones along the trenches bounding many continental margins, together with many other geophysical (e.g. gravimetric) and geological observations, showed how the oceanic crust could disappear into the mantle, providing the mechanism to balance the extension of the ocean basins with shortening along its margins.

All this evidence, both from the ocean floor and from the continental margins, made it clear around 1965 that continental drift was feasible and the theory of plate tectonics, which was defined in a series of papers between 1965 and 1967, was born, with all its extraordinary explanatory and predictive power. The theory revolutionized the Earth sciences, explaining a diverse range of geological phenomena and their implications in other studies such as paleogeography and paleobiology.

Continental Drift

In the late nineteenth and early twentieth centuries, geologists assumed that the Earth's major features were fixed, and that most geologic features such as basin development and mountain ranges could be explained by vertical crustal movement, described in what is called the geosynclinal theory. Generally, this was placed in the context of a contracting planet Earth due to heat loss in the course of a relatively short geological time.

It was observed as early as 1596 that the opposite coasts of the Atlantic Ocean—or, more precisely, the edges of the continental shelves—have similar shapes and seem to have once fitted together.

Since that time many theories were proposed to explain this apparent complementarity, but the assumption of a solid Earth made these various proposals difficult to accept.

The discovery of radioactivity and its associated heating properties in 1895 prompted a re-examination of the apparent age of the Earth. This had previously been estimated by its cooling rate and assumption the Earth's surface radiated like a black body. Those calculations had implied that, even if it started at red heat, the Earth would have dropped to its present temperature in a few tens of millions of years. Armed with the knowledge of a new heat source, scientists realized that the Earth would be much older, and that its core was still sufficiently hot to be liquid.

By 1915, after having published a first article in 1912, Alfred Wegener was making serious arguments for the idea of continental he noted how the east coast of South America and the west coast of Africa looked as if they were once attached. Wegener was not the first to note this (Abraham Ortelius, Antonio Snider-Pellegrini, Eduard Suess, Roberto Mantovani and Frank Bursley Taylor preceded him just to mention a few), but he was the first to marshal significant fossil and paleo-topographical and climatological evidence to support this simple observation (and was supported in this by researchers such as Alex du Toit). Furthermore, when the rock strata of the margins of separate continents are very similar it suggests that these rocks were formed in the same way, implying that they were joined initially. For instance, parts of Scotland and Ireland contain rocks very similar to those found in Newfoundland and New Brunswick. Furthermore, the Caledonian Mountains of Europe and parts of the Appalachian Mountains of North America are very similar in structure and lithology.

However, his ideas were not taken seriously by many geologists, who pointed out that there was no apparent mechanism for continental drift. Specifically, they did not see how continental rock could plow through the much denser rock that makes up oceanic crust. Wegener could not explain the force that drove continental drift, and his vindication did not come until after his death in 1930.

Floating Continents, Paleomagnetism and Seismicity Zones

As it was observed early that although granite existed on continents, seafloor seemed to be composed of denser basalt, the prevailing concept during the first half of the twentieth century was that there were two types of crust, named "sial" (continental type crust) and "sima" (oceanic type crust). Furthermore, it was supposed that a static shell of strata was present under the continents. It therefore looked apparent that a layer of basalt (sial) underlies the continental rocks.

However, based on abnormalities in plumb line deflection by the Andes in Peru, Pierre Bouguer had deduced that less-dense mountains must have a downward projection into the denser layer underneath. The concept that mountains had "roots" was confirmed by George B. Airy a hundred years later, during study of Himalayan gravitation, and seismic studies detected corresponding density variations. Therefore, by the mid-1950s, the question remained unresolved as to whether mountain roots were clenched in surrounding basalt or were floating on it like an iceberg.

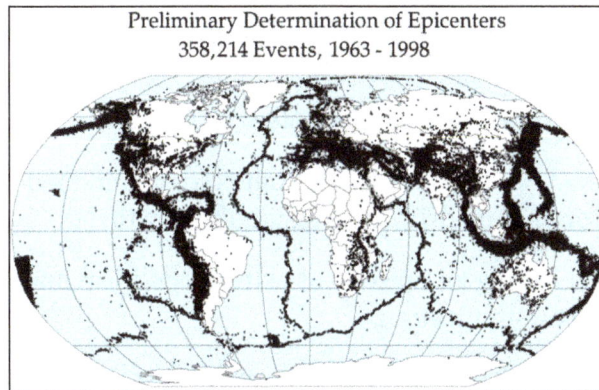

Global earthquake epicenters, 1963–1998.

During the 20th century, improvements in and greater use of seismic instruments such as seismographs enabled scientists to learn that earthquakes tend to be concentrated in specific areas, most notably along the oceanic trenches and spreading ridges. By the late 1920s, seismologists were beginning to identify several prominent earthquake zones parallel to the trenches that typically were inclined 40–60° from the horizontal and extended several hundred kilometers into the Earth. These zones later became known as Wadati-Benioff zones, or simply Benioff zones, in honor of the seismologists who first recognized them, Kiyoo Wadati of Japan and Hugo Benioff of the United States. The study of global seismicity greatly advanced in the 1960s with the establishment of the Worldwide Standardized Seismograph Network (WWSSN) to monitor the compliance of the 1963 treaty banning above-ground testing of nuclear weapons. The much improved data from the WWSSN instruments allowed seismologists to map precisely the zones of earthquake concentration worldwide.

Meanwhile, debates developed around the phenomena of polar wander. Since the early debates of continental drift, scientists had discussed and used evidence that polar drift had occurred because continents seemed to have moved through different climatic zones during the past. Furthermore, paleomagnetic data had shown that the magnetic pole had also shifted during time. Reasoning in an opposite way, the continents might have shifted and rotated, while the pole remained relatively fixed. The first time the evidence of magnetic polar wander was used to support the movements of continents was in a paper by Keith Runcorn in 1956, and successive papers by him and his students Ted Irving (who was actually the first to be convinced of the fact that paleomagnetism supported continental drift) and Ken creer.

This was immediately followed by a symposium in Tasmania in March 1956. In this symposium, the evidence was used in the theory of an expansion of the global crust. In this hypothesis the shifting of the continents can be simply explained by a large increase in size of the Earth since its formation. However, this was unsatisfactory because its supporters could offer no convincing mechanism to produce a significant expansion of the Earth. Certainly there is no evidence that the moon has expanded in the past 3 billion years; other work would soon show that the evidence was equally in support of continental drift on a globe with a stable radius.

During the thirties up to the late fifties, works by Vening-Meinesz, Holmes, Umbgrove, and numerous others outlined concepts that were close or nearly identical to modern plate tectonics theory. In particular, the English geologist Arthur Holmes proposed in 1920 that plate junctions might lie beneath the sea, and in 1928 that convection currents within the mantle might be the driving force. Often, these contributions are forgotten because:

- At the time, continental drift was not accepted.

- Some of these ideas were discussed in the context of abandoned fixistic ideas of a deforming globe without continental drift or an expanding Earth.

- They were published during an episode of extreme political and economic instability that hampered scientific communication.

- Many were published by European scientists and at first not mentioned or given little credit in the papers on sea floor spreading published by the American researchers in the 1960s.

Mid-oceanic Ridge Spreading and Convection

In 1947, a team of scientists led by Maurice Ewing utilizing the Woods Hole Oceanographic Institution's research vessel Atlantis and an array of instruments, confirmed the existence of a rise in the central Atlantic Ocean, and found that the floor of the seabed beneath the layer of sediments consisted of basalt, not the granite which is the main constituent of continents. They also found that the oceanic crust was much thinner than continental crust. All these new findings raised important and intriguing questions.

The new data that had been collected on the ocean basins also showed particular characteristics regarding the bathymetry. One of the major outcomes of these datasets was that all along the globe, a system of mid-oceanic ridges was detected. An important conclusion was that along this system, new ocean floor was being created, which led to the concept of the "Great Global Rift." This was described in the crucial paper of Bruce Heezen (1960), which would trigger a real revolution in thinking. A profound consequence of seafloor spreading is that new crust was, and still is, being continually created along the oceanic ridges. Therefore, Heezen advocated the so-called "expanding Earth" hypothesis of S. Warren Carey. So, still the question remained: how can new crust be continuously added along the oceanic ridges without increasing the size of the Earth? In reality, this question had been solved already by numerous scientists during the forties and the fifties, like Arthur Holmes, Vening-Meinesz, Coates and many others: The crust in excess disappeared along what were called the oceanic trenches, where so-called "subduction" occurred. Therefore, when various scientists during the early sixties started to reason on the data at their disposal regarding the ocean floor, the pieces of the theory quickly fell into place.

The question particularly intrigued Harry Hammond Hess, a Princeton University geologist and a Naval Reserve Rear Admiral, and Robert S. Dietz, a scientist with the U.S. Coast and Geodetic Survey who first coined the term seafloor spreading. Dietz and Hess were among the small handful who really understood the broad implications of sea floor spreading and how it would eventually agree with the, at that time, unconventional and unaccepted ideas of continental drift and the elegant and mobilistic models proposed by previous workers like Holmes.

In the same year, Robert R. Coats of the U.S. Geological Survey described the main features of island arc subduction in the Aleutian Islands. His paper, though little noted (and even ridiculed) at the time, has since been called "seminal" and "prescient." In reality, it actually shows that the work by the European scientists on island arcs and mountain belts performed and published during the 1930s up until the 1950s was applied and appreciated also in the United States.

If the Earth's crust was expanding along the oceanic ridges, Hess and Dietz reasoned like Holmes and others before them, it must be shrinking elsewhere. Hess followed Heezen, suggesting that new oceanic crust continuously spreads away from the ridges in a conveyor belt–like motion. And, using the mobilistic concepts developed before, he correctly concluded that many millions of years later, the oceanic crust eventually descends along the continental margins where oceanic trenches—very deep, narrow canyons—are formed, e.g. along the rim of the Pacific Ocean basin. The important step Hess made was that convection currents would be the driving force in this process, arriving at the same conclusions as Holmes had decades before with the only difference that the thinning of the ocean crust was performed using Heezen's mechanism of spreading along the ridges. Hess therefore concluded that the Atlantic Ocean was expanding while the Pacific Ocean was shrinking. As old oceanic crust is "consumed" in the trenches (like Holmes and others, he thought this was done by thickening of the continental lithosphere, not, as now understood, by under thrusting at a larger scale of the oceanic crust itself into the mantle), new magma rises and erupts along the spreading ridges to form new crust. In effect, the ocean basins are perpetually being "recycled," with the creation of new crust and the destruction of old oceanic lithosphere occurring simultaneously. Thus, the new mobilistic concepts neatly explained why the Earth does not get bigger with sea floor spreading, why there is so little sediment accumulation on the ocean floor, and why oceanic rocks are much younger than continental rocks.

Magnetic Striping

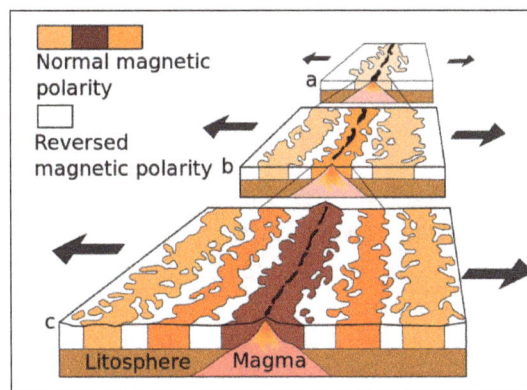

Seafloor magnetic striping.

Beginning in the 1950s, scientists like Victor Vacquier, using magnetic instruments (magnetometers) adapted from airborne devices developed during World War II to detect submarines, began recognizing odd magnetic variations across the ocean floor. This finding, though unexpected, was not entirely surprising because it was known that basalt—the iron-rich, volcanic rock making up the ocean floor—contains a strongly magnetic mineral (magnetite) and can locally distort compass readings. This distortion was recognized by Icelandic mariners as early as the late eighteenth century. More important, because the presence of magnetite gives the basalt measurable magnetic properties, these newly discovered magnetic variations provided another means to study the deep ocean floor. When newly formed rock cools, such magnetic materials recorded the Earth's magnetic field at the time.

A demonstration of magnetic striping
(The darker the color is, the closer it is to normal polarity).

As more and more of the seafloor was mapped during the 1950s, the magnetic variations turned out not to be random or isolated occurrences, but instead revealed recognizable patterns. When these magnetic patterns were mapped over a wide region, the ocean floor showed a zebra-like pattern: one stripe with normal polarity and the adjoining stripe with reversed polarity. The overall pattern, defined by these alternating bands of normally and reversely polarized rock, became known as magnetic striping, and was published by Ron G. Mason and co-workers in 1961, who did not find, though, an explanation for these data in terms of sea floor spreading, like Vine, Matthews and Morley a few years later.

The discovery of magnetic striping called for an explanation. In the early 1960s scientists such as Heezen, Hess and Dietz had begun to theorise that mid-ocean ridges mark structurally weak zones where the ocean floor was being ripped in two lengthwise along the ridge crest (see the previous paragraph). New magma from deep within the Earth rises easily through these weak zones and eventually erupts along the crest of the ridges to create new oceanic crust. This process, at first denominated the "conveyer belt hypothesis" and later called seafloor spreading, operating over many millions of years continues to form new ocean floor all across the 50,000 km-long system of mid-ocean ridges.

Only four years after the maps with the "zebra pattern" of magnetic stripes were published, the link between sea floor spreading and these patterns was correctly placed, independently by Lawrence Morley, and by Fred Vine and Drummond Matthews, in 1963, now called the Vine-Matthews-Morley hypothesis. This hypothesis linked these patterns to geomagnetic reversals and was supported by several lines of evidence:

1. The stripes are symmetrical around the crests of the mid-ocean ridges; at or near the crest of the ridge, the rocks are very young, and they become progressively older away from the ridge crest;

2. The youngest rocks at the ridge crest always- have present-day (normal) polarity;

3. Stripes of rock parallel to the ridge crest alternate in magnetic polarity (normal-reversed-normal, etc.), suggesting that they were formed during different epochs documenting the (already known from independent studies) normal and reversal episodes of the Earth's magnetic field.

By explaining both the zebra-like magnetic striping and the construction of the mid-ocean ridge system, the seafloor spreading hypothesis (SFS) quickly gained converts and represented another major advance in the development of the plate-tectonics theory. Furthermore, the oceanic crust now came to be appreciated as a natural "tape recording" of the history of the geomagnetic field reversals

(GMFR) of the Earth's magnetic field. Today, extensive studies are dedicated to the calibration of the normal-reversal patterns in the oceanic crust on one hand and known timescales derived from the dating of basalt layers in sedimentary sequences (magneto stratigraphy) on the other, to arrive at estimates of past spreading rates and plate reconstructions.

Refining of the Theory

After all these considerations, Plate Tectonics (or, as it was initially called "New Global Tectonics") became quickly accepted in the scientific world, and numerous papers followed that defined the concepts:

- In 1965, Tuzo Wilson who had been a promoter of the sea floor spreading hypothesis and continental drift from the very beginning added the concept of transform faults to the model, completing the classes of fault types necessary to make the mobility of the plates on the globe work out.

- A symposium on continental drift was held at the Royal Society of London in 1965 which must be regarded as the official start of the acceptance of plate tectonics by the scientific community, and which abstracts are issued as Blacket, Bullard & Runcorn. In this symposium, Edward Bullard and co-workers showed with a computer calculation how the continents along both sides of the Atlantic would best fit to close the ocean, which became known as the famous "Bullard's Fit".

- In 1966 Wilson published the paper that referred to previous plate tectonic reconstructions, introducing the concept of what is now known as the "Wilson Cycle."

- In 1967, at the American Geophysical Union's meeting, W. Jason Morgan proposed that the Earth's surface consists of 12 rigid plates that move relative to each other.

- Two months later, Xavier Le Pichon published a complete model based on 6 major plates with their relative motions, which marked the final acceptance by the scientific community of plate tectonics.

- In the same year, McKenzie and Parker independently presented a model similar to Morgan's using translations and rotations on a sphere to define the plate motions.

Principles of Plate Tectonics

In essence, plate-tectonic theory is elegantly simple. Earth's surface layer, 50 to 100 km (30 to 60 miles) thick, is rigid and is composed of a set of large and small plates. Together, these plates constitute the lithosphere, from the Greek lithos, meaning "rock." The lithosphere rests on and slides over an underlying partially molten (and thus weaker but generally denser) layer of plastic partially molten rock known as the asthenosphere, from the Greek asthenos, meaning "weak." Plate movement is possible because the lithosphere-asthenosphere boundary is a zone of detachment. As the lithospheric plates move across Earth's surface, driven by forces as yet not fully understood, they interact along their boundaries, diverging, converging, or slipping past each other. While the interiors of the plates are presumed to remain essentially undeformed, plate boundaries are the sites of many of the principal processes that shape the terrestrial surface, including earthquakes, volcanism, and orogeny (that is, formation of mountain ranges).

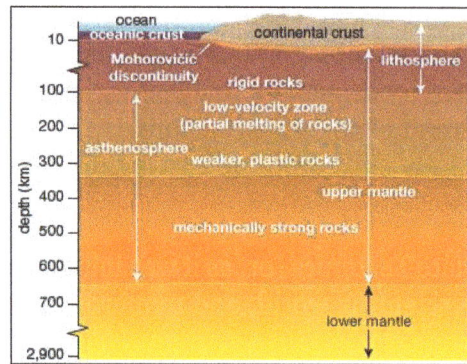

A cross section of Earth's outer layers, from the crust through the lower mantle.

The process of plate tectonics may be driven by convection in Earth's mantle, the pull of heavy old pieces of crust into the mantle, or some combination of both.

Earth's Layers

Knowledge of Earth's interior is derived primarily from analysis of the seismic waves that propagate through Earth as a result of earthquakes. Depending on the material they travel through, the waves may either speed up, slow down, bend, or even stop if they cannot penetrate the material they encounter.

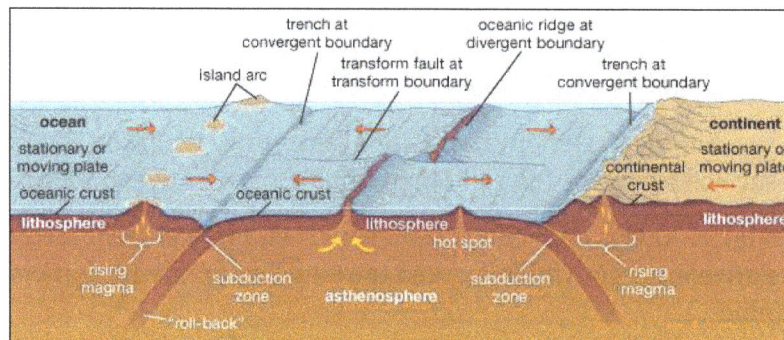

Crustal generation and destruction.

Collectively, these studies show that Earth can be internally divided into layers on the basis of either gradual or abrupt variations in chemical and physical properties. Chemically, Earth can be divided into three layers. A relatively thin crust, which typically varies from a few kilometres to 40 km (about 25 miles) in thickness, sits on top of the mantle. (In some places, Earth's crust may be up to 70 km [40 miles] thick.) The mantle is much thicker than the crust; it contains 83 percent of Earth's volume and continues to a depth of 2,900 km (1,800 miles). Beneath the mantle is the core, which extends to the centre of Earth, some 6,370 km (nearly 4,000 miles) below the surface. Geologists maintain that the core is made up primarily of metallic iron accompanied by smaller amounts of nickel, cobalt, and lighter elements, such as carbon and sulfur.

There are two types of crust, continental and oceanic, which differ in their composition and thickness. The distribution of these crustal types broadly coincides with the division into continents and ocean basins, although continental shelves, which are submerged, are underlain by continental crust. The continents have a crust that is broadly granitic in composition and, with a density of about 2.7 grams per cubic cm (0.098 pound per cubic inch), is somewhat lighter than oceanic crust, which is basaltic (i.e., richer in iron and magnesium than granite) in composition and has a density of about 2.9 to 3 grams per cubic cm (0.1 to 0.11 pound per cubic inch). Continental crust is typically 40 km (25 miles) thick, while oceanic crust is much thinner, averaging about 6 km (4 miles) in thickness. These crustal rocks both sit on top of the mantle, which is ultramafic in composition (i.e., very rich in magnesium and iron-bearing silicate minerals). The boundary between the crust (continental or oceanic) and the underlying mantle is known as the Mohorovičić discontinuity (also called Moho), which is named for its discoverer, Croatian seismologist Andrija Mohorovičić. The Moho is clearly defined by seismic studies, which detect an acceleration in seismic waves as they pass from the crust into the denser mantle. The boundary between the mantle and the core is also clearly defined by seismic studies, which suggest that the outer part of the core is a liquid.

The effect of the different densities of lithospheric rock can be seen in the different average elevations of continental and oceanic crust. The less-dense continental crust has greater buoyancy, causing it to float much higher in the mantle. Its average elevation above sea level is 840 metres (2,750 feet), while the average depth of oceanic crust is 3,790 metres (12,400 feet). This density difference creates two principal levels of Earth's surface.

The lithosphere itself includes all the crust as well as the upper part of the mantle (i.e., the region directly beneath the Moho), which is also rigid. However, as temperatures increase with depth, the heat causes mantle rocks to lose their rigidity. This process begins at about 100 km (60 miles) below the surface. This change occurs within the mantle and defines the base of the lithosphere and the top of the asthenosphere. This upper portion of the mantle, which is known as the lithospheric mantle, has an average density of about 3.3 grams per cubic cm (0.12 pound per cubic inch). The asthenosphere, which sits directly below the lithospheric mantle, is thought to be slightly denser at 3.4–4.4 grams per cubic cm (0.12–0.16 pound per cubic inch).

In contrast, the rocks in the asthenosphere are weaker, because they are close to their melting temperatures. As a result, seismic waves slow as they enter the asthenosphere. With increasing depth, however, the greater pressure from the weight of the rocks above causes the mantle to become gradually stronger, and seismic waves increase in velocity, a defining characteristic of the lower mantle. The lower mantle is more or less solid, but the region is also very hot, and thus the rocks can flow very slowly (a process known as creep).

During the late 20th and early 21st centuries, scientific understanding of the deep mantle was greatly enhanced by high-resolution seismological studies combined with numerical modeling and laboratory experiments that mimicked conditions near the core-mantle boundary. Collectively, these studies revealed that the deep mantle is highly heterogeneous and that the layer may play a fundamental role in driving Earth's plates.

At a depth of about 2,900 km (1,800 miles), the lower mantle gives way to Earth's outer core, which is made up of a liquid rich in iron and nickel. At a depth of about 5,100 km (3,200 miles), the outer core transitions to the inner core. Although it has a higher temperature than the outer core, the inner core is solid because of the tremendous pressures that exist near Earth's centre. Earth's inner core is divided into the outer-inner core (OIC) and the inner-inner core (IIC), which differ from one another with respect to the polarity of their iron crystals. The polarity of the iron crystals of the OIC is oriented in a north-south direction, whereas that of the IIC is oriented east-west.

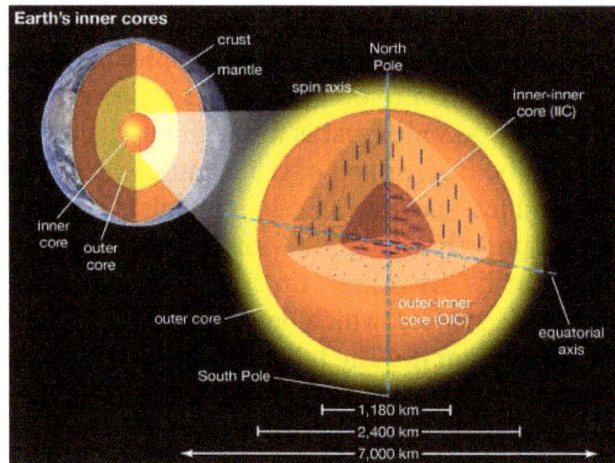

Earth's core.

Plate Boundaries

Lithospheric plates are much thicker than oceanic or continental crust. Their boundaries do not usually coincide with those between oceans and continents, and their behaviour is only partly influenced by whether they carry oceans, continents, or both. The Pacific Plate, for example, is entirely oceanic, whereas the North American Plate is capped by continental crust in the west (the North American continent) and by oceanic crust in the east and extends under the Atlantic Ocean as far as the Mid-Atlantic Ridge.

In a simplified example of plate motion shown in the figure, movement of plate A to the left relative to plates B and C results in several types of simultaneous interactions along the plate boundaries. At the rear, plates A and B move apart, or diverge, resulting in extension and the formation of a divergent margin. At the front, plates A and B overlap, or converge, resulting in compression and the formation of a convergent margin. Along the sides, the plates slide past one another, a process called shear. As these zones of shear link other plate boundaries to one another, they are called transform faults.

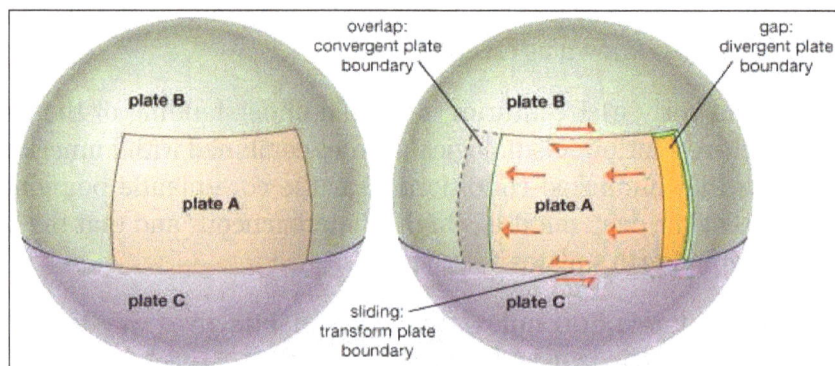

Theoretical diagram showing the effects of an advancing tectonic plate on other adjacent, but stationary, tectonic plates.

At the advancing edge of plate A, the overlap with plate B creates a convergent boundary. In contrast, the gap left behind the trailing edge of plate A forms a divergent boundary with plate B. As plate A slides past portions of both plate B and plate C, transform boundaries develop.

Divergent Margins

As plates move apart at a divergent plate boundary, the release of pressure produces partial melting of the underlying mantle. This molten material, known as magma, is basaltic in composition and is buoyant. As a result, it wells up from below and cools close to the surface to generate new crust. Because new crust is formed, divergent margins are also called constructive margins.

Continental Rifting

Upwelling of magma causes the overlying lithosphere to uplift and stretch. (Whether magmatism [the formation of igneous rock from magma] initiates the rifting or whether rifting decompresses the mantle and initiates magmatism is a matter of significant debate.) If the diverging plates are capped by continental crust, fractures develop that are invaded by the ascending magma, prying the continents farther apart. Settling of the continental blocks creates a rift valley, such as the present-day East African Rift Valley. As the rift continues to widen, the continental crust becomes progressively thinner until separation of the plates is achieved and a new ocean is created. The ascending partial melt cools and crystallizes to form new crust. Because the partial melt is basaltic in composition, the new crust is oceanic, and an ocean ridge develops along the site of the former continental drift. Consequently, diverging plate boundaries, even if they originate within continents, eventually come to lie in ocean basins of their own making.

Seafloor Spreading

As upwelling of magma continues, the plates continue to diverge, a process known as seafloor spreading. Samples collected from the ocean floor show that the age of oceanic crust increases with distance from the spreading centre—important evidence in favour of this process. These age data also allow the rate of seafloor spreading to be determined, and they show that rates vary from about 0.1 cm (0.04 inch) per year to 17 cm (6.7 inches) per year. Seafloor-spreading rates are much more rapid in the Pacific Ocean than in the Atlantic and Indian oceans. At spreading rates of about 15 cm (6 inches) per year, the entire crust beneath the Pacific Ocean (about 15,000 km [9,300 miles] wide) could be produced in 100 million years.

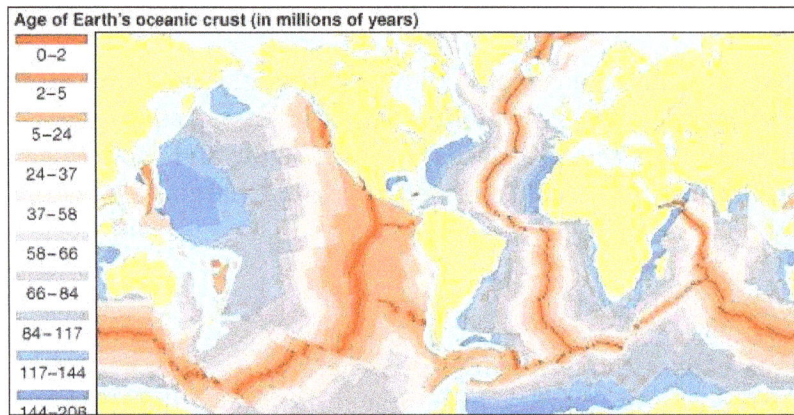

Age of Earth's oceanic crust.

Divergence and creation of oceanic crust are accompanied by much volcanic activity and by many shallow earthquakes as the crust repeatedly rifts, heals, and rifts again. Brittle earthquake-prone rocks occur only in the shallow crust. Deep earthquakes, in contrast, occur less frequently, due to the high heat flow in the mantle rock. These regions of oceanic crust are swollen with heat and so are elevated by 2 to 3 km (1.2 to 1.9 miles) above the surrounding seafloor. The elevated topography results in a feedback scenario in which the resulting gravitational force pushes the crust apart, allowing new magma to well up from below, which in turn sustains the elevated topography. Its summits are typically 1 to 5 km (0.6 to 3.1 miles) below the ocean surface. On a global scale, these ridges form an interconnected system of undersea "mountains" that are about 65,000 km (40,000 miles) in length and are called oceanic ridges.

Convergent Margins

Given that Earth is constant in volume, the continuous formation of Earth's new crust produces an excess that must be balanced by destruction of crust elsewhere. This is accomplished at convergent plate boundaries, also known as destructive plate boundaries, where one plate descends at an angle—that is, is subducted—beneath the other.

Because oceanic crust cools as it ages, it eventually becomes denser than the underlying asthenosphere, and so it has a tendency to subduct, or dive under, adjacent continental plates or younger sections of oceanic crust. The life span of the oceanic crust is prolonged by its rigidity, but eventually this resistance is overcome. Experiments show that the subducted oceanic lithosphere is denser than the surrounding mantle to a depth of at least 600 km (about 400 miles).

The mechanisms responsible for initiating subduction zones are controversial. During the late 20th and early 21st centuries, evidence emerged supporting the notion that subduction zones preferentially initiate along preexisting fractures (such as transform faults) in the oceanic crust. Irrespective of the exact mechanism, the geologic record indicates that the resistance to subduction is overcome eventually.

Where two oceanic plates meet, the older, denser plate is preferentially subducted beneath the younger, warmer one. Where one of the plate margins is oceanic and the other is continental, the greater buoyancy of continental crust prevents it from sinking, and the oceanic plate is preferentially subducted. Continents are preferentially preserved in this manner relative to oceanic crust, which is continuously recycled into the mantle. This explains why ocean floor rocks are generally less than 200 million years old whereas the oldest continental rocks are more than 4 billion years old. Before the middle of the 20th century, most geoscientists maintained that continental crust was too buoyant to be subducted. However, it later became clear that slivers of continental crust adjacent to the deep-sea trench, as well as sediments deposited in the trench, may be dragged down the subduction zone. The recycling of this material is detected in the chemistry of volcanoes that erupt above the subduction zone.

Two plates carrying continental crust collide when the oceanic lithosphere between them has been eliminated. Eventually, subduction ceases and towering mountain ranges, such as the Himalayas, are created.

Because the plates form an integrated system, it is not necessary that new crust formed at any given divergent boundary be completely compensated at the nearest subduction zone, as long as the total amount of crust generated equals that destroyed.

Subduction Zones

The subduction process involves the descent into the mantle of a slab of cold hydrated oceanic lithosphere about 100 km (60 miles) thick that carries a relatively thin cap of oceanic sediments. The path of descent is defined by numerous earthquakes along a plane that is typically inclined between 30° and 60° into the mantle and is called the Wadati-Benioff zone, for Japanese seismologist Kiyoo Wadati and American seismologist Hugo Benioff, who pioneered its study. Between 10 and 20 percent of the subduction zones that dominate the circum-Pacific ocean basin are subhorizontal (that is, they subduct at angles between 0° and 20°). The factors that govern the dip of the subduction zone are not fully understood, but they probably include the age and thickness of the subducting oceanic lithosphere and the rate of plate convergence.

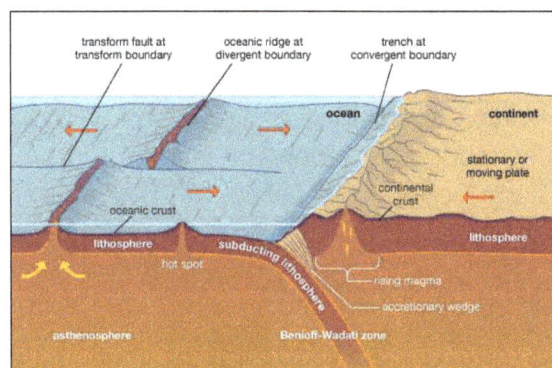

Subducting tectonic plate.

Most, but not all, earthquakes in this planar dipping zone result from compression, and the seismic activity extends 300 to 700 km (200 to 400 miles) below the surface, implying that the subducted crust retains some rigidity to this depth. At greater depths the subducted plate is partially recycled into the mantle.

The site of subduction is marked by a deep trench, between 5 and 11 km (3 and 7 miles) deep, that is produced by frictional drag between the plates as the descending plate bends before it subducts. The overriding plate scrapes sediments and elevated portions of ocean floor off the upper crust of the lower plate, creating a zone of highly deformed rocks within the trench that becomes attached, or accreted, to the overriding plate. This chaotic mixture is known as an accretionary wedge.

The rocks in the subduction zone experience high pressures but relatively low temperatures, an effect of the descent of the cold oceanic slab. Under these conditions the rocks recrystallize, or metamorphose, to form a suite of rocks known as blueschists, named for the diagnostic blue mineral called glaucophane, which is stable only at the high pressures and low temperatures found in subduction zones. Eclogites, which consist of high-pressure minerals such as red garnet (pyrope) and omphacite (pyroxene), form. The formation of eclogite from blueschist is accompanied by a significant increase in density and has been recognized as an important additional factor that facilitates the subduction process.

Island Arcs

When the downward-moving slab reaches a depth of about 100 km (60 miles), it gets sufficiently warm to drive off its most volatile components, thereby stimulating partial melting of mantle in the plate above the subduction zone (known as the mantle wedge). Melting in the mantle wedge produces magma, which is predominantly basaltic in composition. This magma rises to the surface and gives birth to a line of volcanoes in the overriding plate, known as a volcanic arc, typically a few hundred kilometres behind the oceanic trench. The distance between the trench and the arc, known as the arc-trench gap, depends on the angle of subduction. Steeper subduction zones have relatively narrow arc-trench gaps. A basin may form within this region, known as a fore-arc basin, and may be filled with sediments derived from the volcanic arc or with remains of oceanic crust.

If both plates are oceanic, as in the western Pacific Ocean, the volcanoes form a curved line of islands, known as an island arc that is parallel to the trench, as in the case of the Mariana Islands and the adjacent Mariana Trench. If one plate is continental, the volcanoes form inland, as they do in the Andes of western South America. Though the process of magma generation is similar, the ascending magma may change its composition as it rises through the thick lid of continental crust, or it may provide sufficient heat to melt the crust. In either case, the composition of the volcanic mountains formed tends to be more silicon-rich and iron- and magnesium-poor relative to the volcanic rocks produced by ocean-ocean convergence.

Back-arc Basins

Where both converging plates are oceanic, the margin of the older oceanic crust will be subducted because older oceanic crust is colder and therefore more dense. As the dense slab collapses into the asthenosphere, however, it also may "roll back" oceanward and cause extension in the overlying plate. This results in a process known as back-arc spreading, in which a basin opens up behind the island arc. The crust behind the arc becomes progressively thinner, and the decompression of the underlying mantle causes the crust to melt, initiating seafloor-spreading processes, such as melting and the production of basalt; these processes are similar to those that occur at ocean ridges. The geochemistry of the basalts produced at back-arc basins superficially resembles that of basalts produced at ocean ridges, but subtle trace element analyses can detect the influence of a nearby subducted slab.

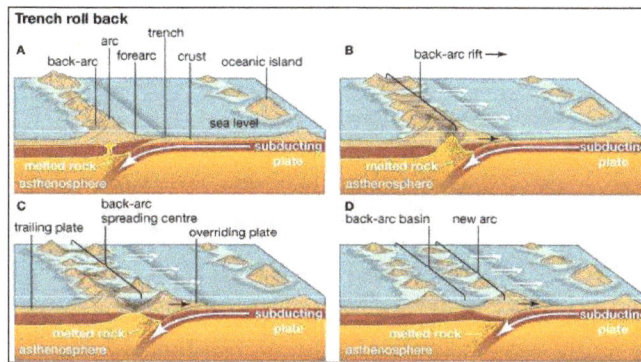

The trench "roll back" process of back-arc basin formation.

This style of subduction predominates in the western Pacific Ocean, in which a number of back-arc basins separate several island arcs from Asia. Examples include the Mariana Islands, the Kuril Islands, and the main islands of Japan. However, if the rate of convergence increases or if anomalously thick oceanic crust (possibly caused by rising mantle plume activity) is conveyed into the subduction zone, the slab may flatten. Such flattening causes the back-arc basin to close, resulting in deformation, metamorphism, and even melting of the strata deposited in the basin.

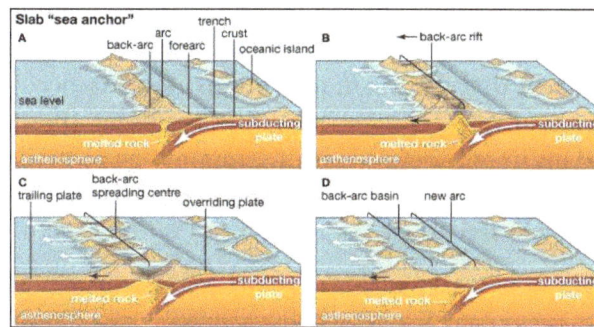

The slab "sea anchor" process of back-arc basin formation.

Mountain Building

If the rate of subduction in an ocean basin exceeds the rate at which the crust is formed at oceanic ridges, a convergent margin forms as the ocean initially contracts. This process can lead to collision between the approaching continents, which eventually terminates subduction. Mountain building can occur in a number of ways at a convergent margin: mountains may rise as a consequence of the subduction process itself, by the accretion of small crustal fragments (which, along with linear island chains and oceanic ridges, are known as terranes), or by the collision of two large continents.

Many mountain belts were developed by a combination of these processes. For example, the Cordilleran mountain belt of North America—which includes the Rocky Mountains as well as the Cascades, the Sierra Nevada, and other mountain ranges near the Pacific coast—developed by a combination of subduction and terrane accretion. As continental collisions are usually preceded by a long history of subduction and terrane accretion, many mountain belts record all three processes. Over the past 70 million years the subduction of the Neo-Tethys Sea, a wedge-shaped body of water that was located between Gondwana and Laurasia, led to the accretion of terranes along the margins of Laurasia, followed by continental collisions beginning about 30 million years ago between Africa and Europe and between India and Asia. These collisions culminated in the formation of the Alps and the Himalayas.

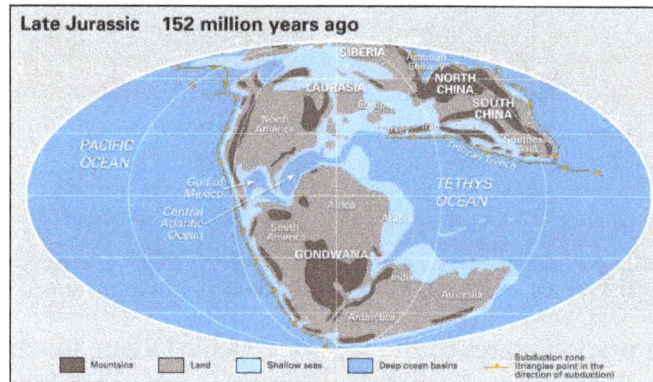

Jurassic paleogeography.

Mountains by Subduction

Mountain building by subduction is classically demonstrated in the Andes Mountains of South America. Subduction results in voluminous magmatism in the mantle and crust overlying the subduction zone, and, therefore, the rocks in this region are warm and weak. Although subduction is a long-term process, the uplift that results in mountains tends to occur in discrete episodes and may reflect intervals of stronger plate convergence that squeezes the thermally weakened crust upward. For example, rapid uplift of the Andes approximately 25 million years ago is evidenced by a reversal in the flow of the Amazon River from its ancestral path toward the Pacific Ocean to its modern path, which empties into the Atlantic Ocean.

In addition, models have indicated that the episodic opening and closing of back-arc basins have been the major factors in mountain-building processes, which have influenced the plate-tectonic evolution of the western Pacific for at least the past 500 million years.

Mountains by Terrane Accretion

As the ocean contracts by subduction, elevated regions within the ocean basin—terranes—are transported toward the subduction zone, where they are scraped off the descending plate and added—accreted—to the continental margin. Since the late Devonian and early Carboniferous periods, some 360 million years ago, subduction beneath the western margin of North America has resulted in several collisions with terranes. The piecemeal addition of these accreted terranes has added an average of 600 km (400 miles) in width along the western margin of the North American continent, and the collisions have resulted in important pulses of mountain building.

During these accretionary events, small sections of the oceanic crust may break away from the subducting slab as it descends. Instead of being subducted, these slices are thrust over the overriding plate and are said to be obducted. Where this occurs, rare slices of ocean crust, known as ophiolites, are preserved on land. They provide a valuable natural laboratory for studying the composition and character of the oceanic crust and the mechanisms of their emplacement and preservation on land. A classic example is the Coast Range ophiolite of California, which is one of the most extensive ophiolite terranes in North America. These ophiolite deposits run from the

Klamath Mountains in northern California southward to the Diablo Range in central California. This oceanic crust likely formed during the middle of the Jurassic Period, roughly 170 million years ago, in an extensional regime within either a back-arc or a forearc basin. In the late Mesozoic, it was accreted to the western North American continental margin.

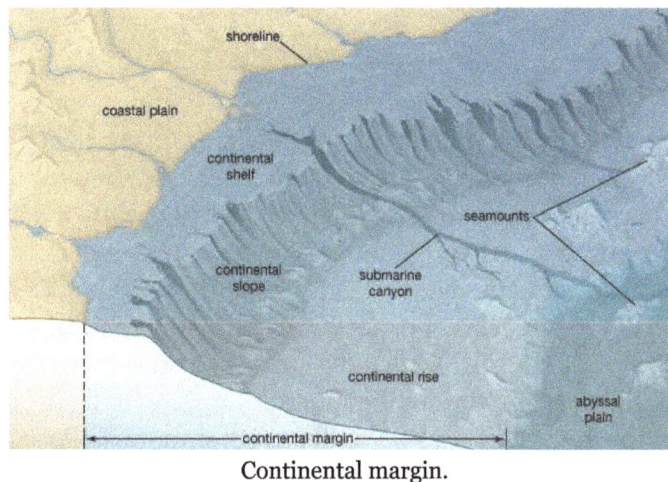

Continental margin.

Because preservation of oceanic crust is rare, the recognition of ophiolite complexes is very important in tectonic analyses. Until the mid-1980s, ophiolites were thought to represent vestiges of the main oceanic tract, but geochemical analyses have clearly indicated that most ophiolites form near volcanic arcs, such as in back-arc basins characterized by subduction roll-back (the collapse of the subducting plate that causes the extension of the overlying plate). The recognition of ophiolite complexes is very important in tectonic analysis, because they provide insights into the generation of magmatism in oceanic domains, as well as their complex relationships with subduction processes.

Mountains by Continental Collision

Continental collision involves the forced convergence of two buoyant plate margins that results in neither continent being subducted to any appreciable extent. A complex sequence of events ensues that compels one continent to override the other. These processes result in crustal thickening and intense deformation that forces the crust skyward to form huge mountains with crustal roots that extend as deep as 80 km (about 50 miles) relative to Earth's surface, in accordance with the principles of isostasy.

The subducted slab still has a tendency to sink and may become detached and founder (submerge) into the mantle. The crustal root undergoes metamorphic reactions that result in a significant increase in density and may cause the root to also founder into the mantle. Both processes result in a significant injection of heat from the compensatory upwelling of asthenosphere, which is an important contribution to the rise of the mountains.

Continental collisions produce lofty landlocked mountain ranges such as the Himalayas. Much later, after these ranges have been largely leveled by erosion, it is possible that the original contact, or suture, may be exposed.

The balance between creation and destruction on a global scale is demonstrated by the expansion of the Atlantic Ocean by seafloor spreading over the past 200 million years, compensated by the contraction of the Pacific Ocean, and the consumption of an entire ocean between India and Asia (the Tethys Sea). The northward migration of India led to collision with Asia some 40 million years ago. Since that time India has advanced a further 2,000 km (1,250 miles) beneath Asia, pushing up the Himalayas and forming the Plateau of Tibet. Pinned against stable Siberia, China and Indochina were pushed sideways, resulting in strong seismic activity thousands of kilometres from the site of the contin9ental collision.

Transform Faults

Along the third type of plate boundary, two plates move laterally and pass each other along giant fractures in Earth's crust. Transform faults are so named because they are linked to other types of plate boundaries. The majority of transform faults link the offset segments of oceanic ridges. However, transform faults also occur between plate margins with continental crust—for example, the San Andreas Fault in California and the North Anatolian fault system in Turkey. These boundaries are conservative because plate interaction occurs without creating or destroying crust. Because the only motion along these faults is the sliding of plates past each other, the horizontal direction along the fault surface must parallel the direction of plate motion. The fault surfaces are rarely smooth, and pressure may build up when the plates on either side temporarily lock. This buildup of stress may be suddenly released in the form of an earthquake.

Section of the San Andreas Fault in the Carrizo Plain.

Many transform faults in the Atlantic Ocean are the continuation of major faults in adjacent continents, which suggests that the orientation of these faults might be inherited from preexisting weaknesses in continental crust during the earliest stages of the development of oceanic crust. On the other hand, transform faults may themselves be reactivated, and recent geodynamic models suggest that they are favourable environments for the initiation of subduction zones.

Hotspots

Although most of Earth's volcanic activity is concentrated along or adjacent to plate boundaries, there are some important exceptions in which this activity occurs within plates. Linear chains of islands, thousands of kilometres in length, that occur far from plate boundaries are the most notable examples. These island chains record a typical sequence of decreasing elevation along the chain, from volcanic island to fringing reef to atoll and finally to submerged seamount. An active volcano usually exists at one end of an island chain, with progressively older extinct volcanoes occurring along the rest of the chain. Canadian geophysicist J. Tuzo Wilson and American geophysicist W. Jason Morgan explained such topographic features as the result of hotspots.

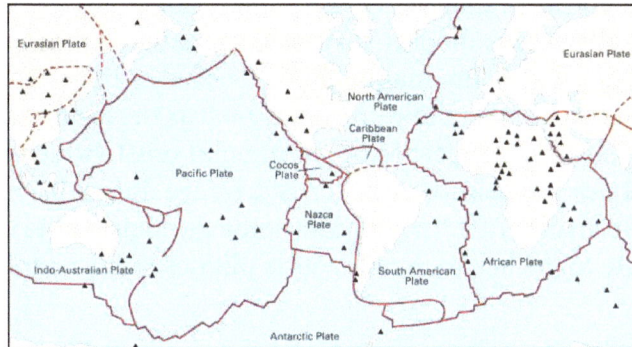

The principal tectonic plates that make up Earth's lithosphere. Also located are several dozen hot spots where plumes of hot mantle material are upwelling beneath the plates.

The number of these hotspots is uncertain (estimates range from 20 to 120), but most occur within a plate rather than at a plate boundary. Hotspots are thought to be the surface expression of giant plumes of heat, termed mantle plumes, that ascend from deep within the mantle, possibly from the core-mantle boundary, some 2,900 km (1,800 miles) below the surface. These plumes are thought to be stationary relative to the lithospheric plates that move over them. A volcano builds upon the surface of a plate directly above the plume. As the plate moves on, however, the volcano is separated from its underlying magma source and becomes extinct. Extinct volcanoes are eroded as they cool and subside to form fringing reefs and atolls, and eventually they sink below the surface of the sea to form a seamount. At the same time, a new active volcano forms directly above the mantle plume.

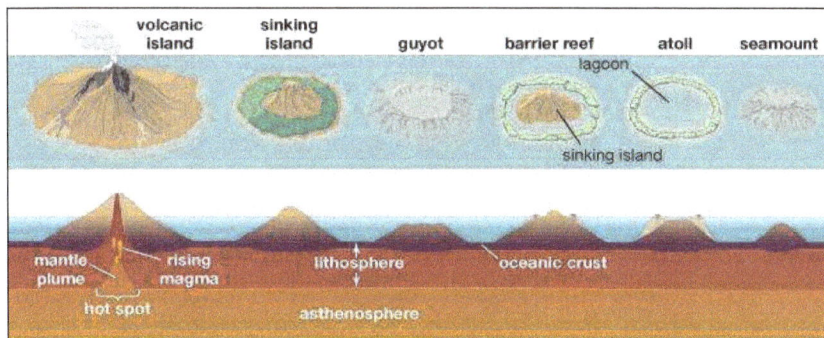

Diagram depicting the process of atoll formation. Atolls are formed from the remnant parts of sinking volcanic islands.

The best example of this process is preserved in the Hawaiian-Emperor seamount chain. The plume is presently situated beneath Hawaii, and a linear chain of islands, atolls, and seamounts extends 3,500 km (2,200 miles) northwest to Midway and a further 2,500 km (1,500 miles) north-north-west to the Aleutian Trench. The age at which volcanism became extinct along this chain gets progressively older with increasing distance from Hawaii—critical evidence that supports this theory. Hotspot volcanism is not restricted to the ocean basins; it also occurs within continents, as in the case of Yellowstone National Park in western North America.

Measurements suggest that hotspots may move relative to one another, a situation not predicted by the classical model, which describes the movement of lithospheric plates over stationary mantle plumes. This has led to challenges to this classic model. Furthermore, the relationship between hotspots and plumes is hotly debated. Proponents of the classical model maintain that these discrepancies are due to the effects of mantle circulation as the plumes ascend, a process called the mantle wind. Data from alternative models suggest that many plumes are not deep-rooted. Instead, they provide evidence that many mantle plumes occur as linear chains that inject magma into fractures, result from relatively shallow processes such as the localized presence of water-rich mantle, stem from the insulating properties of continental crust (which leads to the buildup of trapped mantle heat and decompression of the crust), or are due to instabilities in the interface between continental and oceanic crust. In addition, some geologists note that many geologic processes that others attribute to the behaviour of mantle plumes may be explained by other forces.

Plate Motion

Euler's Contributions

In the 18th century, Swiss mathematician Leonhard Euler showed that the movement of a rigid body across the surface of a sphere can be described as a rotation (or turning) around an axis that goes through the centre of the sphere, known as the axis of rotation. The location of this axis bears no relationship to Earth's spin axis. The point of emergence of the axis through the surface of the sphere is known as the pole of rotation. This theorem of spherical geometry provides an elegant way to define the motion of the lithospheric plates across Earth's surface. Therefore, the relative motion of two rigid plates may be described as rotations around a common axis, known as the axis of spreading. Application of the theorem requires that the plates not be internally deformed—a requirement not absolutely adhered to but one that appears to be a reasonable approximation of what actually happens. Application of this theorem permits the mathematical reconstruction of past plate configurations.

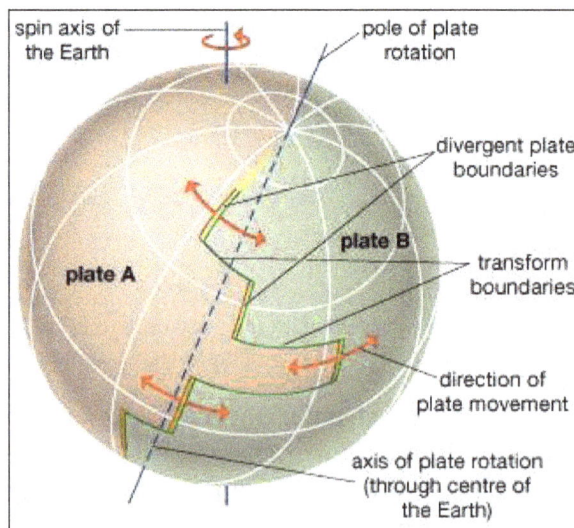

Theoretical depiction of the movement of tectonic plates across Earth's surface.

Movement on a sphere of two plates, A and B, can be described as a rotation around a common pole. Circles around that pole correspond to the orientation of transform faults (that is, single lines in the horizontal that connect to divergent plate boundaries, marked by double lines, in the vertical).

Because all plates form a closed system, all movements can be defined by dealing with them two at a time. The joint pole of rotation of two plates can be determined from their transform boundaries, which are by definition parallel to the direction of motion. Thus, the plates move along transform faults, whose trace defines circles of latitude perpendicular to the axis of spreading, and so form small circles around the pole of rotation. A geometric necessity of this theorem—that lines perpendicular to the transform faults converge on the pole of rotation—is confirmed by measurements. According to this theorem, the rate of plate motion should be slowest near the pole of rotation and increase progressively to a maximum rate along fractures with a 90° angle to it. This relationship is also confirmed by accurate measurements of seafloor-spreading rates.

Past Plate Movements

Plate tectonics involves the movements of Earth's lithospheric plates relative to one another over the planet's weak asthenosphere. This activity changes the positions of all plates with respect to Earth's spin axis and the Equator. To determine the true geographic positions of the plates in the past, investigators have to define their motions, not only relative to each other but also relative to this independent frame of reference. Hotspots, as classically interpreted, provide an example of such a reference frame, assuming they are the sources of plumes that originate within the deep mantle and have relatively fixed positions over time. If this assumption is valid, the motion of the lithosphere above these plumes can be deduced. The hotspot island chains serve this purpose, their trends providing the direction of motion of a plate. The speed of the plate can be inferred from the increase in age of the volcanoes along the chain relative to the distance between the islands.

Earth scientists are able to accurately reconstruct the positions and movements of plates for the past 150 million to 200 million years because they have the oceanic crust record to provide them with plate speeds and direction of movement. However, since older oceanic crust is continuously consumed to make room for new crust, this kind of evidence is not available for earlier intervals of geologic time, making it necessary for investigators to turn to other, less-precise techniques.

Types of Plates

Transform Boundaries

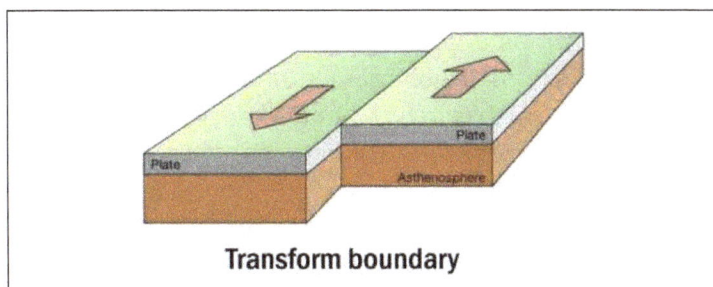

Transform boundary

Transform boundaries (Conservative) occur where two lithospheric plates slide, or perhaps more accurately, grind past each other along transform faults, where plates are neither created nor destroyed. The relative motion of the two plates is either sinistral (left side toward the observer) or dextral (right side toward the observer). Transform faults occur across a spreading center. Strong earthquakes can occur along a fault. The San Andreas Fault in California is an example of a transform boundary exhibiting dextral motion.

Divergent Boundaries

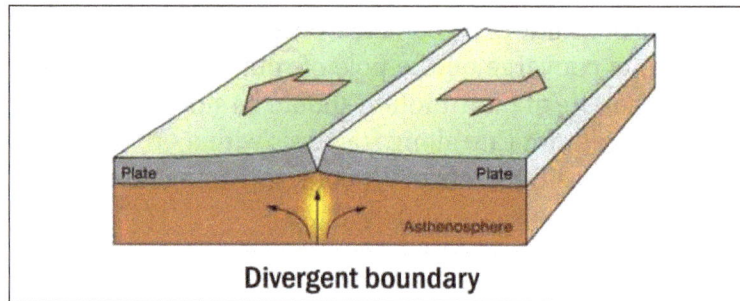

Divergent boundary

Divergent boundaries (Constructive) occur where two plates slide apart from each other. At zones of ocean-to-ocean rifting, divergent boundaries form by seafloor spreading, allowing for the formation of new ocean basin. As the ocean plate splits, the ridge forms at the spreading center, the ocean basin expands, and finally, the plate area increases causing many small volcanoes and shallow earthquakes. At zones of continent-to-continent rifting, divergent boundaries may cause new ocean basin to form as the continent splits, spreads, the central rift collapses, and ocean fills the basin. Active zones of Mid-ocean ridges (e.g., Mid-Atlantic Ridge and East Pacific Rise), and continent-to-continent rifting (such as Africa's East African Rift and Valley, Red Sea) are examples of divergent boundaries.

Convergent Boundaries

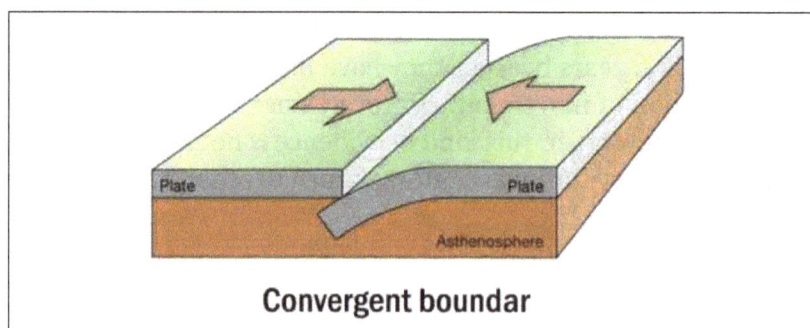

Convergent boundar

Convergent boundary, also known as a destructive plate boundary, is a region of active deformation where two or more tectonic plates or fragments of the lithosphere near the end of their life cycle. This is in contrast to a constructive plate boundary (also known as a mid-ocean ridge or spreading center). As a result of pressure, friction, and plate material melting in the mantle, earthquakes and volcanoes are common near destructive boundaries, where subduction zones or an area of continental collision (depending on the nature of the plates involved) occurs. The subducting plate in a subduction zone is normally oceanic crust, and moves beneath the other plate, which can be made of either oceanic or continental crust. During collisions between two continental plates, large mountain ranges, such as the Himalayas are formed. In other regions, a divergent boundary or transform faults may be present.

References

- Planets: phys.org, Retrieved 2 February, 2019

- How-was-earth-formed: space.com, Retrieved 11 May, 2019

- Evolution-of-earth: scientificamerican.com, Retrieved 20 August, 2019

- Source-of-information-about-the-internal-structure: preservearticles.com, Retrieved 15 January, 2019

- The-inner-structure-of-the-Earth: researchgate.net, Retrieved 10 April, 2019

- The-composition-and-structure-of-earth, geophysical: lumenlearning.com, Retrieved 21 July, 2019

- Crust: nationalgeographic.org, Retrieved 25 January, 2019

- Mantle: nationalgeographic.org, Retrieved 5 March, 2019

- Core: nationalgeographic.org, Retrieved 7 May, 2019

- Continental-drift-theory: thoughtco.com, Retrieved 16 April, 2019

- Evidence-in-support-of-Continental-Drift, continental-drift-theory-tectonics-evidences-continental-drift: pmfias.com, Retrieved 6 March, 2019

- Seafloor-spreading, science: britannica.com, Retrieved 26 June, 2019

- Outcome-theory-of-plate-tectonics: lumenlearning.com, Retrieved 28 February, 2019

- Plate-tectonics, science: britannica.com, Retrieved 8 June, 2019

Chapter 3

Mineral and Rocks

A solid compound which occurs naturally is known as a mineral. Rocks are solid masses which are comprised of minerals. Some of the properties of minerals, on the basis of which they can be classified are color, hardness and tenacity. All these diverse properties of minerals and rocks have been carefully analyzed in this chapter.

Minerals are solid substances that are present in nature and can be made of one element or more elements combined together (chemical compounds).

Gold, Silver and carbon are elements that form minerals on their own. They are called native elements. Instead, ordinary kitchen salt is a chemical compound that is called rock salt, which is a mineral formed of sodium and chlorine ions. Atoms, ions and molecules that form a mineral are present in the space in a tidy way and according to well-defined geometrical shapes, which are called crystal lattices. The structure of the crystal lattice defines the shape of the crystal as we see it. For example, rock salt or kitchen salt is a mineral formed of cubic-shaped crystals. Its crystal lattice has the same shape and consists of sodium and chlorine ions that are present in the space in alternate order.

The order of atoms in the space and the way they combine with each other determine the way a mineral can laminate or exfoliate. Lamination is the property that some materials have to break according to their geometrical shape. Its chemical composition also determines the colour of the crystal, such as the yellow colour for the topaz, red for ruby, purple for amethyst quartz. Another characteristic of minerals is their hardness, which is their resistance to scratches. Hardness is classified by numbers (from 1 to 10), according to the Moh's scale.

At the beginning of the scale there are very soft minerals that can be scratched with a nail, such as talc, chalk and calcite. At the end of the scale there is the diamond, which is the hardest mineral in nature.

To meet the definition of "mineral" used by most geologists, a substance must meet five requirements:

- Naturally occurring

- Inorganic

- Solid

- Definite chemical composition

- Ordered internal structure

"Naturally occurring" means that people did not make it. Steel is not a mineral because it is an alloy produced by people. "Inorganic" means that the substance is not made by an organism. Wood and pearls are made by organisms and thus are not minerals. "Solid" means that it is not a liquid or a gas at standard temperature and pressure.

"Definite chemical composition" means that all occurrences of that mineral have a chemical composition that varies within a specific limited range. For example: the mineral halite (known as "rock salt" when it is mined) has a chemical composition of NaCl. It is made up of an equal number of atoms of sodium and chlorine.

"Ordered internal structure" means that the atoms in a mineral are arranged in a systematic and repeating pattern. Halite is composed of an equal ratio of sodium and chlorine atoms arranged in a cubic pattern.

Major Elements

Almost 99% of the minerals making up the Earth's crust are made up of just eight elements. Most of these elements are found combined with other elements as compounds. Minerals are elements or compounds that occur naturally in the Earth's crust. Rocks are mixtures formed of minerals. Just as elements are the building blocks of minerals, so minerals form the building blocks of rocks.

Table: The elements in the Earth's crust.

Element name	Symbol	Percentage by weight of the earth's crust
Oxygen	O	47
Silicon	Si	28
Aluminium	Al	8
Iron	Fe	5
Calcium	Ca	3.5
Sodium	Na	3
Potassium	K	2.5
Magnesium	Mg	2
All other elements		1

The 'other' elements include copper, uranium, gold and silver. Although they are comparatively rare, these are very important to mankind.

Physical Characteristics

The Physical properties of minerals are used by Mineralogists to help determine the identity of a specimen. Some of the tests can be performed easily in the field, while others require laboratory equipment. For the beginning student of geology, there are a number of simple tests that can be used with a good degree of accuracy. The list of tests is in a suggested order, progressing from simple experimentation and observation to more complicated either in procedure or concept.

Properties of Minerals

The following physical properties of minerals can be easily used to identify a mineral:

1. Color
2. Streak
3. Hardness
4. Cleavage or Fracture
5. Crystalline Structure
6. Diaphaneity or Amount of Transparency
7. Tenacity
8. Magnetism
9. Luster
10. Odor
11. Taste
12. Specific Gravity

Properties of Minerals

Color

Most minerals have a distinctive color that can be used for identification. In opaque minerals, the color tends to be more consistent, so learning the colors associated with these minerals can be very helpful in identification. Translucent to transparent minerals have a much more varied degree of color due to the presence of trace minerals. Therefore, color alone is not reliable as a single identifying characteristic.

Streak

Streak is the color of the mineral in powdered form. Streak shows the true color of the mineral. In large solid form, trace minerals can change the color appearance of a mineral by reflecting the light in a certain way. Trace minerals have little influence on the reflection of the small powdery particles of the streak.

The streak of metallic minerals tends to appear dark because the small particles of the streak absorb the light hitting them. Non-metallic particles tend to reflect most of the light so they appear lighter in color or almost white.

Because streak is a more accurate illustration of the mineral's color, streak is a more reliable property of minerals than color for identification.

Hardness

Hardness is one of the better properties of minerals to use for identifying a mineral. Hardness is a measure of the mineral's resistance to scratching. The Mohs scale is a set of 10 minerals whose hardness is known. The softest mineral, talc, has a Mohs scale rating of one. Diamond is the hardest mineral and has a rating of ten. Softer minerals can be scratched by harder minerals because the forces that hold the crystals together are weaker and can be broken by the harder mineral.

The following is a listing of the minerals of the Mohs scale and their rating:

1. Talc

2. Gypsum

3. Calcite

4. Fluorite

5. Apatite

6. Orthoclase Feldspar

7. Quartz

8. Topaz

9. Corundum

10. Diamond

Cleavage and Fracture

Minerals tend to break along lines or smooth surfaces when hit sharply. Different minerals break in different ways showing different types of cleavage.

Cleavage is defined using two sets of criteria. The first set of criteria describes how easily the cleavage is obtained. Cleavage is considered perfect if it is easily obtained and the cleavage planes are easily distinguished. It is considered good if the cleavage is produced with some difficulty but has obvious cleavage planes. Finally it is considered imperfect if cleavage is obtained with difficulty and some of the planes are difficult to distinguish.

The second set of criteria is the direction of the cleavage surfaces. The names correspond to the shape formed by the cleavage surfaces: Cubic, rhombohedral, octahedral, dodecahedral, basal or prismatic.

Fracture describes the quality of the cleavage surface. Most minerals display either uneven or grainy fracture, conchoidal (curved, shell-like lines) fracture, or hackly (rough, jagged) fracture.

Crystalline Structure

Mineral crystals occur in various shapes and sizes. The particular shape is determined by the arrangement of the atoms, molecules or ions that make up the crystal and how they are joined. This is called the crystal lattice. There are degrees of crystalline structure, in which the fibers of the crystal become increasingly difficult or impossible to see with the naked eye or the use of a hand lens. Microcrystalline and cryptocrystalline structures can only be viewed using high magnification. If there is no crystalline structure, it is called amorphous. However, there are very few amorphous crystals and these are only observed under extremely high magnification.

Transparency or Diaphaneity

Diaphaneity is a mineral's degree of transparency or ability to allow light to pass through it. The degree of transparency may also depend on the thickness of the mineral.

Tenacity

Tenacity is the characteristic that describes how the particles of a mineral hold together or resist separation.

Magnetism

Magnetism is the characteristic that allows a mineral to attract or repel other magnetic materials. It can be difficult to determine the differences between the various types of magnetism, but it is worth knowing that there are distinctions made.

Luster

Luster is the property of minerals that indicates how much the surface of a mineral reflects light. The luster of a mineral is affected by the brilliance of the light used to observe the mineral surface. Luster of a mineral is described in the following terms:

Metallic the mineral is opaque and reflects light as a metal would. Submettalic The mineral is opaque and dull. The mineral is dark colored. Nonmettalic The mineral does not reflect light like a metal.

Nonmetallic minerals are described using modifiers that refer to commonly known qualities.

Waxy the mineral looks like paraffin or wax. Vitreous mineral looks like broken glass. Pearly the mineral appears iridescent, like a pearl. Silky mineral looks fibrous, like silk. Greasy the mineral looks like oil on water.

- Resinous: The mineral looks like hardened tree sap (resin).
- Adamantine: The mineral looks brilliant, like a diamond.

Odor

Most minerals have no odor unless they are acted upon in one of the following ways: moistened, heated, breathed upon, or rubbed.

Taste

Only soluble minerals have a taste, but it is very important that minerals not be placed in the mouth or on the tongue.

Specific Gravity

Specific Gravity of a mineral is a comparison or ratio of the weight of the mineral to the weight of an equal amount of water. The weight of the equal amount of water is found by finding the difference between the weight of the mineral in air and the weight of the mineral in water.

Major Minerals and their Characteristics

Some common minerals and their characteristics are given below:

1. Feldspar: Silicon and oxygen are common elements in all types of feldspar and sodium, potassium, calcium, aluminum etc. are found in specific feldspar variety. Half of the earth's crust is composed of feldspar. It has light cream to salmon pink color. It is used in ceramics and glass making.

2. Quartz: It is one of the most important components of sand and granite. It consists of silica. It is a hard mineral virtually insoluble in water. It is white or colorless and used in radio and radar. It is one of the most important components of granite.

3. Pyroxene: Pyroxene consists of calcium, aluminum, magnesium, iron, and silica. Pyroxene forms 10 percent of the earth's crust. It is commonly found in meteorites. It is in green or black color.

4. Amphibole: Aluminum, calcium, silica, iron, magnesium are the major elements of amphiboles. They form 7 percent of the earth's crust. It is in green or black color and is used in asbestos industry. Hornblende is another form of amphiboles.

5. Mica: It comprises of potassium, aluminum, magnesium, iron, silica etc. It forms 4 percent of the earth's crust. It is commonly found in igneous and metamorphic rocks. It is used in electrical instruments.

6. Olivine: Magnesium, iron, and silica are major elements of olivine. It is used in jewelry. It is usually a greenish crystal, often found in basaltic rocks. Besides these main minerals, other minerals like chlorite, calcite, magnetite, haematite, bauxite, and barite are also present in some quantities in the rocks.

Rocks

Rock is a naturally occuringand coherent aggregate of one or more minerals. Such aggregates constitute the basic unit of which the solid Earth is comprised and typically form recognizable and mappable volumes. Rocks are commonly divided into three major classes according to the processes that resulted in their formation. These classes are (1) igneous rocks, which have solidified from molten material called magma; (2) sedimentary rocks, those consisting of fragments derived from preexisting rocks or of materials precipitated from solutions; and (3) metamorphic rocks, which have been derived from either igneous or sedimentary rocks under conditions that caused changes in mineralogical composition, texture, and internal structure. These three classes, in turn, are subdivided into numerous groups and types on the basis of various factors, the most important of which are chemical, mineralogical, and textural attributes.

Types

Igneous Rock

Igneous rock, are various crystalline or glassy rocks formed by the cooling and solidification of molten earth material. Igneous rocks constitute one of the three principal classes of rocks, the others being metamorphic and sedimentary.

Igneous rocks are formed from the solidification of magma, which is a hot (600 to 1,300 °C, or 1,100 to 2,400 °F) molten or partially molten rock material. Earth is composed predominantly of a large mass of igneous rock with a very thin veneer of weathered material—namely, sedimentary rock. Whereas sedimentary rocks are produced by processes operating mainly at Earth's surface by the disintegration of mostly older igneous rocks, igneous—and metamorphic—rocks are formed by internal processes that cannot be directly observed and that necessitate the use of physical-chemical arguments to deduce their origins. Because of the high temperatures within Earth, the principles of chemical equilibrium are applicable to the study of igneous and metamorphic rocks, with the latter being restricted to those rocks formed without the direct involvement of magma.

Magma is thought to be generated within the plastic asthenosphere (the layer of partially molten rock underlying Earth's crust) at a depth below about 60 kilometres (40 miles). Because magma is less dense than the surrounding solid rocks, it rises toward the surface. It may settle within the crust or erupt at the surface from a volcano as a lava flow. Rocks formed from the cooling and solidification of magma deep within the crust are distinct from those erupted at the surface mainly owing to the differences in physical and chemical conditions prevalent in the two environments. Within Earth's deep crust the temperatures and pressures are much higher than at its surface; consequently, the hot magma cools slowly and crystallizes completely, leaving no trace of the liquid magma. The slow cooling promotes the growth of minerals large enough to be identified visually without the aid of a microscope (called phaneritic, from the Greek phaneros, meaning "visible"). On the other hand, magma erupted at the surface is chilled so quickly that the individual minerals have little or no chance to grow. As a result, the rock is either composed of minerals that can be seen only with the aid of a microscope (called aphanitic, from the Greek aphanēs, meaning "invisible") or contains no minerals at all (in the latter case, the rock is composed of glass, which is a highly viscous liquid). This results in two groups: (1) plutonic intrusive igneous rocks that solidified deep within the crust and (2) volcanic, or extrusive, igneous rocks formed at Earth's surface. Some intrusive rocks, known as subvolcanic, were not formed at great depth but were instead injected near the surface where lower temperatures result in a more rapid cooling process; these tend to be aphanitic and are referred to as hypabyssal intrusive rocks.

The deep-seated plutonic rocks can be exposed at the surface for study only after a long period of denudation or by some tectonic forces that push the crust upward or by a combination of the two conditions. (Denudation is the wearing away of the terrestrial surface by processes including weathering and erosion.) Generally, the intrusive rocks have cross-cutting contacts with the country rocks that they have invaded, and in many cases the country rocks show evidence of having been baked and thermally metamorphosed at these contacts. The exposed intrusive rocks are found in a variety of sizes, from small veinlike injections to massive dome-shaped batholiths, which extend for more than 100 square kilometres (40 square miles) and make up the cores of the great mountain ranges.

Extrusive rocks occur in two forms: (1) as lava flows that flood the land surface much like a river and (2) as fragmented pieces of magma of various sizes (pyroclastic materials), which often are blown through the atmosphere and blanket Earth's surface upon settling. The coarser pyroclastic materials accumulate around the erupting volcano, but the finest pyroclasts can be found as thin layers located hundreds of kilometres from the opening. Most lava flows do not travel far from the volcano, but some low-viscosity flows that erupted from long fissures have accumulated in thick (hundreds of metres) sequences, forming the great plateaus of the world (e.g., the Columbia River plateau of Washington and Oregon and the Deccan plateau in India). Both intrusive and extrusive magmas have played a vital role in the spreading of the ocean basin, in the formation of the oceanic crust, and in the formation of the continental margins. Igneous processes have been active since the onset of the formation of Earth some 4.6 billion years ago. Their emanations have provided the water for the oceans, the gases for the primordial oxygen-free atmosphere, and many valuable mineral deposits.

Composition

Chemical Components

The great majority of the igneous rocks are composed of silicate minerals (meaning that the basic building blocks for the magmas that formed them are made of silicon [Si] and oxygen [O]), but minor occurrences of carbonate-rich igneous rocks are found as well. Indeed, in 1960 a sodium carbonate (Na_2CO_3) lava with only 0.05 weight percent silica (SiO_2) was erupted from Ol Doinyo Lengai, a volcano in northern Tanzania, Africa. Because of the limited occurrence of such carbonate-rich igneous rocks, however, the following discussion will consider the chemistry of silicate rocks only. The major oxides of the rocks generally correlate well with their silica content: those rocks with low silica content are enriched in magnesium oxide (MgO) and iron oxides (FeO, Fe_2O_3, and Fe_3O_4) and are depleted in soda (Na_2O) and potash (K_2O); those with a large amount of silica are depleted in magnesium oxide and iron oxides but are enriched in soda and potash. Both calcium oxide (CaO) and alumina (Al_2O_3) are depleted in the rocks that have a silica content of less than about 45 weight percent, but, above 45 percent, calcium oxide can be as high as 10 percent; this amount decreases gradually as the silica increases. Alumina in rocks that contain more than 45 percent silica is generally above approximately 14 weight percent, with the greatest abundance occurring at an intermediate silica content of about 56 weight percent. Because of the importance of silica content, it has become common practice to use this feature of igneous rocks as a basis for subdividing them into the following groups: silicic or felsic (or acid, an old and discredited but unfortunately entrenched term), rocks having more than 66 percent silica; intermediate, rocks with 55 to 66 percent silica; and subsilicic, rocks containing less than 55 percent silica. The latter may be further divided into two groups: mafic, rocks with 45 to 55 percent silica and ultramafic, those containing less than 45 percent. The subsilicic rocks, enriched as they are in iron (Fe) and magnesium (Mg), are termed femic (from ferrous iron and magnesium), whereas the silicic rocks are referred to as sialic (from silica and aluminum, with which they are enriched) or salic (from silica and aluminum). The terms mafic (from magnesium and ferrous iron) and felsic (feldspar and silica) are used interchangeably with femic and sialic.

The silica content also reflects the mineral composition of the rocks. As the magma cools and begins to crystallize, silica is taken from the magma to be combined with the other cationic oxides to form the silicate minerals. For example, one mole of SiO_2 is combined with one mole of MgO to make the magnesium-rich pyroxene, $MgSiO_3$ (enstatite): $SiO_2 + MgO \rightarrow MgSiO_3$. Two moles of SiO_2 are needed to be combined with one mole each of CaO and Al_2O_3 to make the calcium-rich plagioclase, $CaAl_2Si_2O_8$ (anorthite). However, in a case where magma does not have enough silica relative to the magnesium oxide to produce the pyroxene, the magma will compensate by making a magnesium-olivine (forsterite; Mg_2SiO_4), along with the pyroxene, since the olivine requires only one-half as much silica for every mole of magnesium oxide. On the other hand, a silicic magma may have excess silica such that some will be left after all the silicate minerals were formed from the combination of the oxides; the remaining "free" silica crystallizes as quartz or its polymorphs. The former case usually occurs in subsilicic rocks that characteristically will have silicate minerals like magnesium-olivine, sodium-nepheline ($NaAlSiO_4$, which requires only one mole of silicon for every mole of sodium [Na]), and leucite ($KAlSi_2O_6$, which requires only two moles of silicon to one mole of potassium [K]). These three minerals substitute in part for enstatite, albite ($NaAlSi_3O_8$, requiring three moles of silicon for one mole of sodium), and orthoclase feldspar ($KAlSi_3O_8$, requiring three moles of silicon for one mole of potassium), respectively. Quartz clearly will not be present in these rocks. Minerals such as magnesium-olivine, nepheline, and leucite are termed undersaturated (with respect to silica), and the subsilicic rocks that contain them are termed undersaturated as well. In the case of rocks that have excess silica, the silicic rocks will have quartz and magnesium-pyroxene, which are considered saturated minerals, and the rocks that contain them are termed supersaturated.

Mineralogical Components

The major mineralogical components of igneous rocks can be divided into two groups: felsic (from feldspar and silica) and mafic (from magnesium and ferrous iron). The felsic minerals include quartz, tridymite, cristobalite, feldspars (plagioclase and alkali feldspar), feldspathoids (nepheline and leucite), muscovite, and corundum. Because felsic minerals lack iron and magnesium, they are generally light in colour and consequently are referred to as such or as leucocratic. The mafic minerals include olivine, pyroxenes, amphiboles, and biotites, all of which are dark in colour. Mafic minerals are said to be melanocratic. These terms can be applied to the rocks, depending on the relative proportion of each type of mineral present. In this regard, the term colour index, which refers to the total percentage of the rock occupied by mafic minerals, is useful. Felsic rocks have a colour index of less than 50, while mafic rocks have a colour index above 50. Those rocks that have a colour index above 90 are referred to as ultramafic. These terms are to be used only for the mineralogical content of igneous rocks because they do not necessarily correlate directly with chemical terms. For example, it is common to find a felsic rock composed almost entirely of the mineral plagioclase, but in chemical terms, such a rock is a subsilicic mafic rock. Another example is an igneous rock consisting solely of pyroxene. Mineralogically it would be termed ultramafic, but chemically it is a mafic igneous rock with a silica content of about 50 percent.

The influence of supersaturation and undersaturation on the mineralogy of a rock was noted above. During the crystallization of magmas, supersaturated minerals will not be formed along with undersaturated minerals. Supersaturated minerals include quartz and its polymorphs and a low-calcium orthorhombic pyroxene. These cannot coexist with any of the feldspathoids (e.g., leucite and nepheline) or magnesium-rich olivine. In volcanic rocks that have been quenched (cooled rapidly) such that only a small part of the magma has been crystallized, it is possible to find a forsterite (magnesium-rich olivine) crystal surrounded by a glass that is saturated or supersaturated. In this case, the outer rim of the olivine may be corroded or replaced by a magnesium-rich pyroxene (called a reaction rim). The olivine was the first to be crystallized, but it was in the process of reacting with the saturated magma to form the saturated mineral when an eruption halted the reaction. Had the magma been allowed to crystallize fully, all the forsterite would have been transformed into the magnesium-rich pyroxene and quartz may have been crystallized.

Accessory minerals present in igneous rocks in minor amounts include monazite, allanite, apatite, garnets, ilmenite, magnetite, titanite, spinel, and zircon. Glass may be a major phase in some volcanic rock but, when present, is usually found in minor amounts. Igneous rocks that were exposed to weathering and circulating groundwater have undergone some degree of alteration. Common alteration products are talc or serpentine formed at the expense of olivine, chlorites replacing pyroxene and amphiboles, iron oxides replacing any mafic mineral, clay minerals and epidote formed from the feldspars, and calcite that may be formed at the expense of any calcium-bearing mineral by interaction with a carbon dioxide (CO_2)-bearing solution. Glass is commonly altered to clay minerals and zeolite. In some cases, however, glass has undergone a devitrification process (in which it is transformed into a crystalline material) initiated by reaction of the glass with water or by subsequent reheating. Common products of devitrification include quartz and its polymorphs, alkali feldspar, plagioclase, pyroxene, zeolite, clays, and chlorite.

Textural Features

The texture of an igneous rock normally is defined by the size and form of its constituent mineral grains and by the spatial relationships of individual grains with one another and with any glass that may be present. Texture can be described independently of the entire rock mass, and its geometric characteristics provide valuable insights into the conditions under which the rock was formed.

Crystallinity

Among the most fundamental properties of igneous rocks are crystallinity and granularity, two terms that closely reflect differences in magma composition and the differences between volcanic and various plutonic environments of formation. Crystallinity generally is described in terms of the four categories shown in the table.

Crystallinity categories of igneous rocks:

Crystallinity	Rock term
Entirely crystalline	Holocrystalline
Crystalline material and subordinate glass	Hemicrystalline or hypocrystalline
Glass and subordinate crystalline material	Hemihyaline or hypohyaline
Entirely glassy	Holohyaline or hyaline

Those holocrystalline rocks in which mineral grains can be recognized with the unaided eye are called phanerites, and their texture is called phaneritic. Those with mineral grains so small that their outlines cannot be resolved without the aid of a hand lens or microscope are termed aphanites, and their texture is termed aphanitic. Aphanitic rocks are further described as either microcrystalline or cryptocrystalline, according to whether or not their individual constituents can be resolved under the microscope. The subaphanitic, or hyaline, rocks are referred to as glassy, or vitric, in terms of granularity.

Aphanitic and glassy textures represent relatively rapid cooling of magma and, hence, are found mainly among the volcanic rocks. Slower cooling, either beneath Earth's surface or within very thick masses of lava, promotes the formation of crystals and, under favourable circumstances of magma composition and other factors, their growth to relatively large sizes. The resulting phaneritic rocks are so widespread and so varied that it is convenient to specify their grain size as shown in the table.

Categories of rock grain size		
Terms in common use	General grain size	
	Igneous rocks in general	Pegmatites
Fine-grained	<1 mm	<1 inch
Medium-grained	1–5 mm	1–4 inches
Coarse-grained	5 mm–2 cm	4–12 inches
Very coarse-grained	>2 cm	>12 inches

Granularity

Grain size

The general grain size ordinarily is taken as the average diameter of dominant grains in the rock; for the pegmatites, which are special rocks with extremely large crystals, it can refer to the maximum exposed dimensions of dominant grains. Most aphanitic rocks are characterized by mineral grains less than 0.3 millimetre (0.01 inch) in diameter, and those in which the average grain size is less than 0.1 millimetre (0.004 inch) are commonly described as dense.

Fabric

A major part of rock texture is fabric or pattern, which is a function of the form and outline of its constituent grains, their relative sizes, and their mutual relationships in space. Many specific terms have been employed to shorten the description of rock fabrics, and even the sampling offered here may seem alarmingly extensive. It should be noted, however, that fabric provides some of the most useful clues to the nature and sequence of magmatic crystallization.

The degree to which mineral grains show external crystal faces can be described as euhedral or panidiomorphic (fully crystal-faced), subhedral or hypidiomorphic (partly faced), or anhedral or allotriomorphic (no external crystal faces). Quite apart from the presence or absence of crystal faces, the shape, or habit, of individual mineral grains is described by such terms as equant, tabular, platy, elongate, fibrous, rodlike, lathlike, needlelike, and irregular. A more general contrast can be drawn between grains of equal (equant) and inequal dimensions. Even-grained, or equigranular, rocks are characterized by essential minerals that all exhibit the same order of grain size, but this implied equality need not be taken too literally. For such rocks the combination terms panidiomorphic-granular, hypidiomorphic-granular, and allotriomorphic-granular are applied according to the occurrence of euhedral, subhedral, and anhedral mineral grains within them. Many fine-grained allotriomorphic-granular rocks are more simply termed sugary, saccharoidal, or aplitic.

Rocks that are unevenly grained, or inequigranular, are generally characterized either by a seriate fabric, in which the variation in grain size is gradual and essentially continuous, or by a porphyritic fabric, involving more than one distinct range of grain sizes. Both of these kinds of texture are common. The relatively large crystals in a porphyritic rock ordinarily occur as separate entities, known as phenocrysts, set in a groundmass or matrix of much finer-grained crystalline material or glass. Quite commonly in many volcanic rocks, phenocrysts are aggregated. When this is observed, the term glomeroporphyritic is used to describe the texture, and the aggregate is referred to as a glomerocryst. In some cases, such glomerocrysts are monomineralic, but more commonly they are composed of two or more minerals. Based on chemical composition, texture, and other criteria such as isotopic analysis, it has been demonstrated that some phenocrysts and glomerocrysts were not crystallized from the host magma but rather were accidentally torn from the country rock by the magma as it rose to the surface. When this has occurred, these phenocrysts are referred to as xenocrysts, while the aggregates can be termed xenoliths. The size of phenocrysts is essentially independent of their abundance relative to the groundmass, and they range in external form from euhedral to anhedral. Most of them are best described as subhedral. Because the groundmass constituents span almost the full ranges of crystallinity and granularity, porphyritic fabric is abundantly represented among the phaneritic, aphanitic, and glassy rocks.

The sharp break in grain size between phenocrysts and groundmass reflects a corresponding change in the conditions that affected the crystallizing magma. Thus, the phenocrysts of many rocks probably grew slowly at depth, following which the nourishing magma rose to Earth's surface as lava, cooled much more rapidly, and congealed to form a finer-grained or glassy groundmass. A porphyritic volcanic rock with a glassy groundmass is described as having a vitrophyric texture and the rock can be called a vitrophyre. Other porphyritic rocks may well reflect less drastic shifts in position and perhaps more subtle and complex changes in conditions of temperature, pressure, or crystallization rates. Many phenocrysts could have developed at the points where they now occur, and some may represent systems with two fluid phases, magma and coexisting gas. Appraisals of the composition of phenocrysts, their distribution, and their periods of growth relative to the accompanying groundmass constituents are important to an understanding of many igneous processes.

Important Textural Types

The articulation of mineral grains is described in terms of planar, smoothly curved, sinuous, sutured, interlocked, or irregular surfaces of mutual boundary. The distribution and orientation of mineral grains and of mineral grains and glass are other elements of fabric that can be useful in estimating the conditions and sequence of mineral formation in igneous rocks. The following are only a few of the most important examples:

- Directive textures are produced by the preferred orientation of platy, tabular, or elongate mineral grains to yield grossly planar or linear arrangements; they are generally a result of magmatic flowage.

- Graphic texture refers to the regular intergrowth of two minerals, one of them generally serving as a host and the other appearing on surfaces of the host as striplike or cuneiform units with grossly consistent orientation; the graphic intergrowth of quartz in alkali feldspar is a good example.

- Ophitic texture is the association of lath-shaped euhedral crystals of plagioclase, grouped radially or in an irregular mesh, with surrounding or interstitial large anhedral crystals of pyroxene; it is characteristic of the common rock type known as diabase.

- Poikilitic texture describes the occurrence of one mineral that is irregularly scattered as diversely oriented crystals within much larger host crystals of another mineral.

- Reaction textures occur at the corroded margins of crystals, from the corrosive rimming of crystals of one mineral by finer-grained aggregates of another, or as a result of other features that indicate partial removal of crystalline material by reaction with magma or other fluid.

- Pyroclastic texture results from the explosive fragmentation of volcanic material, including magma (commonly the light, frothy pumice variety and glass fragments called shards), country rock, and phenocrysts. Fragments less than 2 millimetres in size are called ash, and the rock formed of these is called tuff; fragments between 2 and 64 millimetres are lapilli and the rock is lapillistone; fragments greater than 64 millimetres are called bombs if rounded or blocks if angular, and the corresponding rock is termed agglomerate or pyroclastic breccia, respectively.

Commonly, many of these pyroclastic rocks have been formed by dense hot clouds that hug the ground and behave much like a lava flow and hence are given the name pyroclastic flow. Most of these flows are composed of ash-size material; therefore, they are called ash flows and the rocks deposited by them are called ash-flow tuffs. A more general term for rocks deposited by these flows that does not specify size of fragments is ignimbrite. Ash-flow tuffs and other ignimbrites often have zones in which the fragments have been welded. These zones are termed welded tuffs and display a directive planar texture (called eutaxitic) that results from compaction and flattening of pumice fragments. Such pyroclastic flows were responsible for many of the deposits of the eruption of Mount St. Helens in Washington State, U.S., on May 18, 1980. Most eruptions eject fragments that are borne by the wind and deposited subaerially (on the land surface). These deposits are said to be ash-fall tuffs and are recognized by their lamination (formation in thin layers that differ in grain size or composition). They commonly blanket the topography in contrast to the ash-flow deposits, which flow around topographic highs and which are completely unsorted.

- Replacement textures occur where a mineral or mineral aggregate has the external crystal form of a preexisting different mineral (pseudomorphism) or where the juxtaposition of two minerals indicates that one was formed at the expense of the other.

Finally, crystal zoning describes faintly to very well-defined geometric arrangements of portions within individual crystals that differ significantly in composition (or some other property) from adjacent portions; most common are successive shells grouped concentrically about the centres of crystals, presumably reflecting shifts in conditions during crystal growth.

Structural Features

The structure of an igneous rock is normally taken to comprise the mutual relationships of mineral or mineral-glass aggregates that have contrasting textures, along with layering, fractures, and other larger-scale features that transect or bound such aggregates. Structure often can be described only in relation to masses of rock larger than a hand specimen, and most of its individual expressions can be closely correlated with physical conditions that existed when the rock was formed.

Small-scale Structural Features

Among the most widespread structural features of volcanic rocks are the porelike openings left by the escape of gas from the congealing lava. Such openings are called vesicles, and the rocks in which they occur are said to be vesicular. Where the openings lie close together and form a large part of the containing rock, they impart to it a slaglike, or scoriaceous, structure. Their relative abundance is even greater in the type of sialic glassy rock known as pumice, which is essentially a congealed volcanic froth. Most vesicles can be likened to peas or nuts in their ranges of size and shape; those that were formed when the lava was still moving tend to be flattened and drawn out in the direction of flow. Others are cylindrical, pearlike, or more irregular in shape, depending in part on the manner of escape of the gas from the cooling lava; most of the elongate ones occur in subparallel arrangements.

Many vesicles have been partly or completely filled with quartz, chalcedony, opal, calcite, epidote, zeolites, or other minerals. These fillings are known as amygdules, and the rock in which they are present is amygdaloidal. Some are concentrically layered, others also include centrally disposed series of horizontal layers, and still others are featured by central cavities into which well-formed crystals project.

Spherulites are light-coloured subspherical masses that commonly consist of tiny fibres and plates of alkali feldspar radiating outward from a centre. Most range from pinpoint to nut size, but some are as much as several feet in diameter. The relatively large ones tend to be internally complex and to contain concentric shells of feldspar fibres with or without accompanying quartz, tridymite, or glass. Spherulites occur mainly in glassy volcanic rocks; they also are present in some partly or wholly crystalline rocks that include shallow-seated intrusive types. Many evidently are products of rapid crystallization, perhaps at points of gas concentration in the freezing magmas. Others, in contrast, were formed more slowly, by devitrification of volcanic glasses, presumably not long after they congealed and while they were still relatively hot.

Lithophysae, also known as stone bubbles, consist of concentric shells of finely crystalline alkali feldspar separated by empty spaces; thus, they resemble an onion or a newly blooming rose. Commonly associated with spherulites in glassy and partly crystalline volcanic rocks of salic composition, many lithophysae are about the size of walnuts. They have been ascribed to short episodes of rapid crystallization, alternating with periods of gas escape when the open spaces were developed by thrusting the feldspathic shells apart or by contraction associated with cooling. The curving cavities commonly are lined with tiny crystals of quartz, tridymite, feldspar, topaz, or other minerals deposited from the gases.

Some glassy rocks of silicic composition are marked by domains of strongly curved, concentrically disposed fractures that promote breakage into rounded masses of pinhead to walnut size. Because their surfaces often have a pearly or shiny lustre, the name perlite is applied to such rocks. Perlite is most common in glassy silicic rocks that have interacted with water to become hydrated. During the hydration process, water enters the glass, breaking the silicon-oxygen bonds and causing an expansion of the glass structure to form the curved cracks. The extent of hydration of glass, indicating the amount of perlite that has been formed from the glass, depends on the climate and on time. In a given area where the climate is expected to be consistent, the thickness of the hydration of the glass surface has been used by archaeologists to date artifacts such as arrowheads composed of the dark volcanic glass known as obsidian and made by early native Americans.

Numerous structural features of comparably small scale occur among the intrusive rocks; these include miarolitic, orbicular, plumose, and radial structures. Miarolitic rocks are felsic phanerites distinguished by scattered pods or layers, ordinarily several centimetres in maximum thickness, within which their essential minerals are coarser-grained, subhedral to euhedral, and otherwise pegmatitic in texture. Many of these small interior bodies, called miaroles, contain centrally disposed crystal-lined cavities that are known as druses or miarolitic cavities. An internal zonal disposition of minerals also is common, and the most characteristic sequence is alkali feldspar with graphically intergrown quartz, alkali feldspar, and a central filling of quartz. Miarolitic structure probably represents local concentration of gases during very late stages in consolidation of the host rocks.

The term orbicular is applied to rounded, onionlike masses with distinct concentric layering that are distributed in various ways through otherwise normal-appearing phaneritic rocks of silicic to mafic composition. The layers within individual masses are typically thin, irregular, and sharply defined, and each differs from its immediate neighbours in composition or texture. Some layers contain tabular or prismatic mineral grains that are oriented radially with respect to the containing orbicule and, hence, are analogous to spherulitic layers in volcanic rocks. The minerals of most orbicules are the same as those of the enclosing rock, but they are not necessarily present in the same proportions. The concentric structure appears to reflect rhythmic crystallization about specific centres, commonly at early stages in consolidation of the general rock mass.

The normal fabric of some relatively coarse-grained plutonic rocks is interrupted by clusters of crystals with radial grouping but without concentric layering. A characteristic plumelike, spraylike, or rosettelike structure is imparted by the markedly elongate form of the participating crystals or crystal aggregates, which seem to have developed outward from common centres by direct crystallization from magma or by replacement of preexisting solid material.

Large-scale Structural Features

Many kinds of larger-scale features occur among both the intrusive and the extrusive rocks. Most of these are mentioned later in connection with rock occurrence or are discussed in other articles, but several are properly introduced here.

Clastic Structures

These are various features that express the accumulation of fragments or the rupturing and dislocation of solid material. In volcanic environments they generally result from explosive activity or the incorporation of solid fragments by moving lava; as such, they characterize the pyroclastic rocks. Among the plutonic rocks, they appear chiefly as local to very extensive zones of pervasive shearing, dislocation, and granulation, commonly best recognized under the microscope. Those developed prior to final consolidation of the rock are termed protoclastic; those developed after final consolidation, cataclastic.

Flow Structures

These are planar or linear features that result from flowage of magma with or without contained crystals. Various forms of faintly to sharply defined layering and lining typically reflect compositional or textural inhomogeneities, and they often are accentuated by concentrations or preferred orientation of crystals, inclusions, vesicles, spherulites, and other features.

Fractures

These are straight or curving surfaces of rupture directly associated with the formation of a rock or later superimposed upon it. Primary fractures generally can be related to emplacement or to subsequent cooling of the host rock mass. The columnar jointing found in many mafic volcanic rocks is a typical result of contraction upon cooling.

Inclusions

These are rounded to angular masses of solid material enclosed within a rock of recognizably different composition or texture. Those consisting of older material not directly related to that of their host are known as xenoliths, and those representing broken-up and detached older parts of the same igneous body that encloses them are termed cognate xenoliths or autoliths.

Pillow Structures

These are aggregates of ovoid masses, resembling pillows or grain-filled sacks in size and shape, that occur in many basic volcanic rocks. The masses are separated or interconnected, and each has a thick vesicular crust or a thinner and more dense glassy rind. The interiors ordinarily are coarser-grained and less vesicular. Pillow structure is formed by rapid chilling of highly fluid lava in contact with water or water-saturated sediments, accompanied by the development of budlike projections with tough, elastic crusts. As additional lava is fed into each bud, it grows into a pillow and continues to enlarge until rupture of the skin permits escape of fresh lava to form a new bud and a new pillow.

Segregations

These are special types of inclusions that are intimately related to their host rocks and in general are relatively rich in one or more of the host-rock minerals. They range from small pods to extensive layers and from early-stage crystal accumulations formed by gravitational settling in magma to very late-stage concentrations of coarse-grained material developed in place.

Zonal Structures

These are arrangements of rock units with contrasting composition, or texture, in an igneous body, commonly in a broadly concentric pattern. Chilled margins, the fine-grained or glassy edges along the borders of many extrusive and shallow-seated intrusive bodies, represent quenching of magma along contacts with cooler country rock. Other kinds of zones generally reflect fractional crystallization of magma and are useful in tracing courses of magmatic differentiation.

An interesting type of zonal structure is an orbicular configuration that has alternating light and dark repeating bands in an oval arrangement found in some diorites and granodiorites. Pegmatites also often have zonal structures due to fluctuations in fluid composition. This results in "pockets" that may contain gems or other unusual minerals.

Classification of Igneous Rocks

Igneous rocks are classified on the basis of mineralogy, chemistry, and texture. Texture is used to subdivide igneous rocks into two major groups: (1) the plutonic rocks, with mineral grain sizes that are visible to the naked eye, and (2) the volcanic and hypabyssal types, which are usually too fine-grained or glassy for their mineral composition to be observed without the use of a petrographic microscope. Being rather coarsely grained, phaneritic rocks readily lend themselves to a classification based on mineralogy since their individual mineral components can be discerned, but the volcanic rocks are more difficult to classify because either their mineral composition is not visible or the rock has not fully crystallized owing to fast cooling. As a consequence, various methods employ chemical composition as the criterion for volcanic igneous rock classification. A commonly used technique was introduced at the beginning of the 20th century by the American geologists C. Whitman Cross, Joseph P. Iddings, Louis V. Pirsson, and Henry S. Washington. In this method, the mineral composition of the rock is recalculated into a standard set of typically occurring minerals that theoretically could have developed from the complete equilibrium crystallization at low temperatures of a magma of the indicated bulk composition. The calculated hypothetical mineral composition is called the norm, and the minerals constituting the standard set are termed normative minerals, since they are ordinarily found in igneous rocks. The rock under analysis may then be classified according to the calculated proportions of the normative minerals.

Because other methods for calculating the norm have been devised, this original norm is referred to as the CIPW norm after the initials of the four petrologists who devised the system. The norm calculation allows the petrologist studying an aphanitic rock to "see" the mineral assemblage that corresponds well with the actual mineral assemblage of a plutonic rock of the same composition that had crystallized under equilibrium conditions. Moreover, the norm has been shown to have a thermodynamic basis. The concept of silica saturation discussed above is incorporated into the norm, which will show whether a magma of a certain composition is supersaturated, saturated, or undersaturated by the presence or absence of normative minerals such as quartz, orthopyroxene, olivine, and the feldspathoids.

Classification of Plutonic Rocks

A plutonic rock may be classified mineralogically based on the actual proportion of the various minerals of which it is composed (called the mode). In any classification scheme, boundaries between classes are set arbitrarily; however, if the boundaries can be placed closest to natural divisions or gaps between classes, they will seem less random and subjective, and the standards will facilitate universal understanding. In order to set boundaries nearest to the population lows (of constituent minerals) and to achieve an international consensus, a poll among the world's petrologists was conducted and a modal classification for plutonic igneous rocks was devised. Based mainly on this poll, the International Union of Geological Sciences (IUGS) Subcommission on the Systematics of Igneous Rocks in 1973 suggested the use of the modal composition for all plutonic igneous rocks with a colour index less than 90 and for those plutonic ultramafic rocks with a colour index greater than 90.

The plotting of rock modes on these triangular diagrams is simpler than it may appear. If the colour index is less than 90 and quartz (Q) is present, then the three components, Q + A (alkali feldspar) + P (plagioclase), are recalculated from the mode to sum to 100 percent and Figure is used. Each component is represented by the corners of the equilateral triangle, the length of whose sides are divided into 100 equal parts. Any composition plotting at a corner, therefore, has a mode of 100 percent of the corresponding component. Any point on the sides of the triangle represents a mode composed of the two adjacent corner components. For example, a rock with 60 percent Q and 40 percent A will plot on the QA side at a location 60 percent of the distance from A to Q. A rock containing all three components will plot within the triangle. Since the sides of the triangle are divided into 100 parts, a rock having a mode of 20 percent Q and 80 percent A + P (in unknown proportions for the moment) will plot on the line that parallels the AP side and lies 20 percent of the distance toward Q from the side AP. If this same rock has 30 percent P and 50 percent A, the rock mode will plot at the intersection of the 20 percent Q line described above, with a line paralleling the QA side at a distance 30 percent toward P from the QA side. The third intersecting line for the point is necessarily the line paralleling the QP side at 50 percent of the distance from the side QP toward A.

A rock with 25 percent Q, 35 percent P, and 40 percent A plots in the granite field, whereas one with 25 percent Q, 60 percent P, and 15 percent A plots in the granodiorite field. The latter is close to the average composition of the continental crust of Earth. Igneous rocks normally do not exceed about 50 percent quartz, and the feldspathoidal rocks are relatively rare. The most common plutonic rocks are those in fields numbered 3, 4, 5, 8, 9, 10, and 15. These are found in what have been called granite (used in a loose sense) batholiths, which are irregularly shaped large bodies covering an area greater than 100 square kilometres. Batholiths constitute the cores of the great mountain ranges, such as the Rockies in western North America and the Sierra Nevada in California, U.S. Typically these batholiths are composites of smaller intrusions, each of which may display several different rock types. The average composition is close to that of a granodiorite, but in many batholiths the sequence of intrusions progresses from basic to acidic, with gabbro or quartz diorite being emplaced first. In the Sierra Nevada batholith, the dominant rocks are quartz monzonite and granodiorite, with intrusions including quartz diorite in the far western rim and granite in the east. Batholiths contain medium- to coarse-grained rocks with hypidiomorphic-granular texture. The rocks are generally leucocratic; diorites and quartz diorites typically contain less than 30 percent mafic minerals—e.g., hornblende and biotite. Pyroxenes are rare but are more commonly found in the gabbros. Mineralogically the ratio of hornblende to biotite, the colour index, the calcium content of the plagioclase feldspar, and the ratio of plagioclase to alkali feldspar decrease from diorite to quartz diorite to granodiorite and granite. Common accessory minerals include apatite, titanite, and an opaque mineral such as magnetite or ilmenite.

Ideally it would be preferable to use the same modal scheme for volcanic rocks. This is recommended whenever possible; hence, for this purpose, the volcanic or hypabyssal equivalent of the plutonic rocks are listed in parentheses in. It should be noted that the dividing lines and boxes are identical to those for the plutonic rocks.

Classification of Volcanic and Hypabyssal Rocks

Owing to the aphanitic texture of volcanic and hypabyssal rocks, their modes cannot be readily determined; consequently, a chemical classification is widely accepted and employed by most petrologists. One popular scheme is based on the use of both chemical components and normative mineralogy. Because most lay people have little access to analytic facilities that yield igneous rock compositions, only an outline will be presented here in order to provide an appreciation for the classification scheme.

The first major division is based on the alkali (soda + potash) and silica contents, which yield two groups, the subalkaline and alkaline rocks. The subalkaline rocks have two divisions based mainly on the iron content, with the iron-rich group called the tholeiitic series and the iron-poor group called calc-alkalic. The former group is most commonly found along the oceanic ridges and on the ocean floor; the latter group is characteristic of the volcanic regions of the continental margins (convergent, or destructive, plate boundaries. In some magmatic arcs (groups of islands arranged in a curved pattern), notably Japan, both the tholeiitic and calc-alkalic series occur. This is the case, for example, in the volcanoes of northeastern Honshu, the largest of Japan's four main islands, and both series may be found within the same volcano. The alkaline rocks frequently occur on oceanic islands (usually formed during the late stages of magma consolidation after tholeiitic eruptions) and in continental rifts (extensive fractures). Based on the relative proportions of soda and potash, the calc-alkalic series is subdivided into the sodic and potassic series.

Chemically the subalkaline rocks are saturated with respect to silica; consequently, they have normative minerals such as orthopyroxene [$Mg(Fe)_2S_{i2}O_6$] and quartz but lack nepheline and olivine (in the presence of quartz). This chemical property also is reflected in the mode of the basic members that have two pyroxenes, orthopyroxene and augite [$Ca(Mg, Fe)Si_2O_6$], and perhaps quartz. Plagioclase is common in phenocrysts, but it can also occur in holocrystalline rocks in the microcrystalline matrix along with the pyroxenes and an iron–titanium oxide phase. In addition to the differences in iron content between the tholeiitic and calc-alkalic series, the latter has a higher alumina content (16 to 20 percent), and the range in silica content is larger (48 to 75 percent compared to 45 to 63 percent for the former). Hornblende and biotite phenocrysts are common in the calc-alkalic andesites and dacites but are lacking in the tholeiites except as alteration products. The dacites and rhyolites commonly have phenocrysts of plagioclase, alkali feldspar (usually sanidine), and quartz in a glassy matrix. Hornblende and plagioclase phenocrysts are more widespread in dacites than in rhyolites, which have more biotite and alkali feldspar. When occurring near volcanic vents, (openings from which volcanic materials are brought to Earth's surface), basalts and andesites of both series are found as tuffs or agglomerates; otherwise, they typically occur as flows. Dacite and rhyolite occur as flows near vents but are most commonly found as tuffs composed of fragmented pieces of glass, phenocrysts, and rock.

The alkaline rocks typically are chemically undersaturated with respect to silica; hence they lack normative orthopyroxene (i.e., they have only one pyroxene, the calcium-rich augite) and quartz but have normative nepheline. Microscopic examination of the alkali olivine basalts usually reveals phenocrysts with an abundance of olivine, one pyroxene (augite, which is usually titanium-rich), and plagioclase. Nepheline may be seen in the matrix. Trachytes typically are leucocratic with an abundance of feldspars aligned roughly parallel to the direction of the lava flow.

Origin and Distribution

Origin of Magmas

Basaltic magmas that form the oceanic crust of Earth are generated in the asthenosphere at a depth of about 70 kilometres. The mantle rocks located at depths from about 70 to 200 kilometres are believed to exist at temperatures slightly above their melting point, and possibly 1 or 2 percent of the rocks occur in the molten state. As a result, the asthenosphere behaves plastically, and upon penetrating this zone seismic waves experience a slight drop in velocity; this shell came to be known as the low velocity zone. Only after the acceptance of the plate tectonic theory has this zone become known as the asthenosphere. The most common mantle rock within the asthenosphere is peridotite, which is composed predominantly of magnesium-rich olivine, along with lesser amounts of chromium diopside and enstatite and an even smaller quantity of garnet. Peridotite may undergo partial melting to produce magmas with different compositions.

Theories on the generation of basaltic magma mainly attribute its origin to the derivation of heat from within peridotite rather than by some outside source such as the radioactive decay of uranium, thorium, and potassium, which are only of minor consequence. Because of the difference in composition between basalt and peridotite, only a small amount of heat is needed to produce about 3 to at most 25 percent melt. The change in the temperature of Earth as a function of depth, given by the estimated geothermal gradient, and the experimentally based melting curve (solidus) of the peridotite are illustrated in figure. At depth D, the geothermal gradient curve and the solidus of the peridotite have their closest approach, but the peridotite is still solid. Diverse mechanisms have been proposed to explain the cause for the intersection here of the two curves. One theory suggests that a decrease in pressure (equivalent to depth) at constant composition and without loss of heat will cause the peridotite to melt along the curve DS. This is identical to an adiabatic cooling process (one without an overall loss or gain of heat) in which temperature will drop slightly owing to the expansion of the rock that occurs in response to the pressure decrease. The drop in temperature is about 10 times smaller than the drop in temperature along the solidus for the same decrease in pressure. Physically, the peridotite rises to a lesser depth owing to convection in the mantle (the zone below Earth's crust) without any exchange of heat. Melting is initiated when the curve DS intersects the melting curve at point E. As the peridotite continues to rise, it will follow the melting curve, continually producing more melt. This results from the peridotite providing its own heat. To illustrate this, consider the peridotite following the adiabatic curve DS from E to point T where it is (T - F) degrees above the melting curve. Allowing the peridotite to cool at this pressure from T to F releases heat that will be consumed in the melting process. The peridotite can be thought of as making similar but infinitesimally small steps like E to T to F as it moves along the solidus. In this way heat is provided for the melting as the peridotite moves continuously along the solidus.

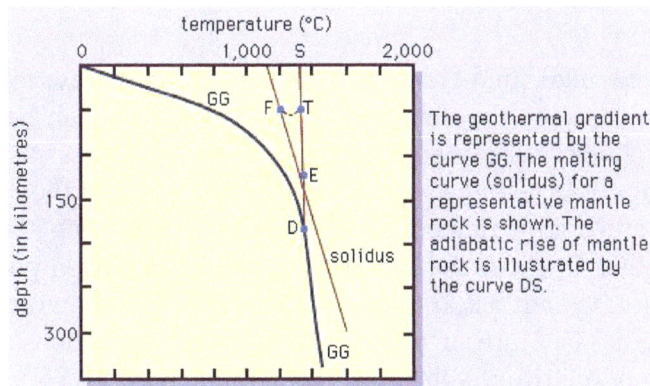

The geothermal gradient is represented by the curve GG. The melting curve (solidus) for a representative mantle rock is shown. The adiabatic rise of mantle rock is illustrated by the curve DS.

A proposed temperature distribution within the Earth.

Granitic, or rhyolitic, magmas and andesitic magmas are generated at convergent plate boundaries where the oceanic lithosphere (the outer layer of Earth composed of the crust and upper mantle) is subducted so that its edge is positioned below the edge of the continental plate or another oceanic plate. Heat will be added to the subducting lithosphere as it moves slowly into the hotter depths of the mantle. The andesitic magma is believed to be generated in the wedge of mantle rock below the crust and above the subducted plate or within the subducted plate itself. The former requires the partial melting of a "wet" peridotite. Experiments conducted at pressures simulating mantle conditions have demonstrated that peridotite will produce andesitic melts during partial melting under hydrous conditions. The latter theory suggests that the subducted basaltic crust is partially melted and may be combined with some subducted oceanic sediments to form andesites. A third theory involves the mixing of basaltic magma that was generated in the mantle with granitic or rhyolitic magma or with crustal rocks. The silicic magmas can be formed by a combination of two processes; the presence of water under pressure lowers the melting temperature by as much as 200 °C (392 °F) and thereby expedites magma generation. At a convergent plate boundary, the lower continental crust is heated to a temperature near its melting point by being pushed downward into hotter regions of the mantle. Basaltic or andesitic magma generated below the crust may accumulate near the Moho, which is a discontinuity that separates Earth's crust from its mantle. As the magma cools, it crystallizes and releases its latent heat of crystallization. This evolved heat is transferred to the lower crustal rocks along with the simple heat released by cooling. If the lower crustal rocks contain some water, their melting temperatures would be lowered and the heating provided by the above processes would possibly be sufficient to partially melt the crustal rocks producing rhyolitic magma.

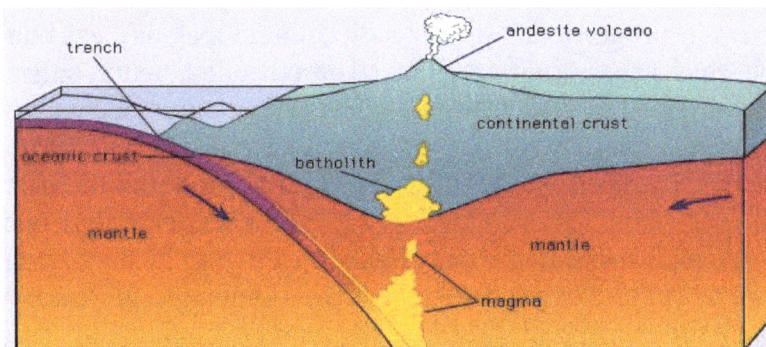

The oceanic plate is shown subducting below the continental crust. Melting of the crust or of the mantle above the subducting plate occurs only at a depth of about 100 kilometers below the Earth's surface.

Nature of Magmas

Magmas are chemically complex fluid systems that differ in many ways from ordinary solutions, in which water is the solvent and the dominant constituent. They can be thought of as mutual solutions, or melts, of rock-forming components that are variously present as simple ions, as complex ions and ionic groups, and as molecules. The most abundant of the simple ions in common magmas are such singly and doubly charged cations as Na^+, K^+, Ca_2^+, Mg_2^+, and Fe_2^+. Because these ions can move about rather freely in the system, they occupy no fixed positions with respect to other ions that are present. In contrast, the smaller and more highly charged cations, notably Si_4^+, Al_3^+, and (to a lesser degree) Fe_3^+, are surrounded or screened by O_2^- ions and other anions (negative ions) to form parts of relatively stable complex ions such as $(SiO_4)_4^-$, $(AlO_4)_5^-$, and $(FeO_6)_9^-$. Simple anions, including F^-, Cl^-, O_2^-, and $(OH)^-$, ordinarily are present in much smaller amounts. Water, hydrochloric acid (HCl), hydrogen fluoride (HF), carbon dioxide (CO_2), and other volatile molecular substances occur as well, generally in equilibrium with ionic forms such as $(OH)^-$, Cl^-, F^-, and $(CO_3)_2^-$.

Because the bond that unites silicon and oxygen is a remarkably strong one, $(SiO_4)_4^-$ ions are stable in magmas even at exceedingly high temperatures. They also tend to join with one another, or polymerize, to form more complex anionic groups, a tendency that is especially great in the more silicic magmas. The joining is accomplished by a sharing of oxygen ions between adjacent silicon ions to form Si-O-Si bridges like those in many silicate and aluminosilicate minerals; in the simplest such case, $(Si_2O_7)_6^-$ ions are the result. Because the $(AlO_4)_5^-$ ions also have a strong tendency to polymerize, most of the large ionic groups in magmas probably contain both silicon and aluminum ions. These groups, which resemble the frameworks of many rock-forming minerals but are geometrically less regular, significantly affect the viscosity and crystallization of magmas.

The viscosity of magmas, which spans an enormous range of values, affects their flow behaviour, the movements of crystals and inclusions of foreign matter within them, the diffusion of materials through them, the growth of crystals from them, and the explosivity of eruptions (when aided by growth of gas bubbles near the surface). Lava flows are thin and rapid for low-viscosity magmas, but thick and slow for viscous flows. Fluid magma promotes the growth of large crystals such as the ones found in pegmatites, but crystal growth is prevented in viscous magmas, which usually are quenched as glass. Highly explosive eruptions such as occurred at Mount St. Helens commonly result from gas bubbles nucleating, growing, and rising in a highly viscous magma. It can be demonstrated thermodynamically that the overpressure (excess rock pressure) developed in growing and rising bubbles is inversely proportional to their radii. In fluid magmas, gas bubbles grow large in size and rise quickly, which causes their pressures to be expended; hence, only a spectacular fountaining of hot lava is observed at the surface. In contrast, a viscous magma prevents the growth of bubbles, so that they will rise slowly while retaining excess pressure; as a result, the associated volcano erupts violently. Energies equivalent to the amount produced by several nuclear bombs are released in such explosions. Viscosity increases greatly with decreasing temperature and less markedly with increasing pressure. It also can be governed in part by the amount and distribution of any solid materials or bubbles of gas present, which both tend to increase viscosity. Finally, it varies considerably among magmas of differing gross composition, mainly because of the differences in the degree of Si-O and Al-O polymerization. Thus, highly silicic magmas generally are more viscous than mafic ones by several orders of magnitude, a difference reflected by contrasts in the eruptive behaviour of rhyolitic and basaltic lavas. Basaltic magmas at 1,100 °C can be at least 100,000 times more viscous than water at room temperature, whereas rhyolitic magmas at 800 °C are at least 10 million times more viscous than room-temperature water. The presence of volatile constituents can markedly increase the fluidity of magmas, even those that are rich in SiO_2. This effect has been attributed to the breaking of Si-O-Si bridges through substitution of ions such as F^- and $(OH)^-$ for shared O_2^- ions in elements of the polymerized groups.

A typical magma can be broadly viewed as an assemblage of relatively large and rather closely packed oxygen ions, among which some cations have considerable mobility; others, such as $Si4+$ and $Al3^+$, tend to occupy positions that are more fixed. The entire system is a dynamic one, however, and even the largest of the Si^- O and Al^- O ion groups are constantly changing form and position as bonds are broken and new ones are established. If the magma quickly loses thermal energy and cools to a glass, these internal movements are sharply restricted, and the various constituents become essentially frozen in position. If cooling is slower, the contained complex ions and polymerized ion groups have time to assume more regular arrangements and to be stabilized by cations of appropriate size, charge, and other properties. Crystalline solids are thereby formed. Their regular internal structure is relatively conserving of space, and so they have somewhat higher specific gravities than the magma from which they were nourished.

Crystallization from Magmas

The Forsterite-cristobalite System

Because magmas are multicomponent solutions, they do not crystallize at a single temperature at a given pressure like water at 0 °C and one atmosphere pressure. Rather, they crystallize over a wide range of temperatures beginning at liquidus temperatures for basaltic magmas as high as 1,150 °C and ending as a complete solid at a low solidus temperature of about 800 °C. During their crystallization at constant pressure, common minerals that make up basaltic magma (e.g., olivine) become unstable at some temperature and react with the liquid to form a more stable phase. In the case of olivine, this phase is pyroxene. This reaction relationship is best illustrated with the use of a phase diagram of a portion of the olivine Mg_2SiO_4 (forsterite) + SiO_2 (cristobalite, a high-temperature form of quartz) binary system at one atmosphere.

Consider a mixture X of two minerals in the proportions 28 percent cristobalite and 72 percent forsterite. At a temperature of 1,601 °C, this mixture is entirely liquid. At temperatures below 1,557 °C, forsterite (F_o) and enstatite (E_n) are stable, but between 1,557 and 1,600 °C, forsterite and the liquid whose composition is represented by L are in equilibrium. At a temperature of 1,570 °C, there is about 7 percent forsterite and 93 percent liquid. As the liquid X cools, it intersects the liquidus freezing curve at a temperature of 1,600 °C, where forsterite begins to crystallize. As the temperature drops further, the liquid follows the liquidus down toward R, the peritectic point (incongruent melting point in a binary system), while it continually crystallizes more forsterite. It should be noted that the liquid composition is becoming enriched in silica, until at R, it has more silica than enstatite. At this point the forsterite reacts with the liquid to yield two moles of $MgSiO_3$ (enstatite) for every mole of Mg_2SiO_4 that combines with one mole of SiO_2 removed from the liquid R.

This can be written as a chemical equation: $Mg_2SiO_4 + SiO_2 \rightleftarrows 2MgSiO_3$. Because SiO_2 is removed from the liquid R, a proportionate amount of enstatite must be crystallized from the liquid to keep its composition at point R. In the case of the starting composition X, which is depleted in SiO_2 relative to enstatite, the peritectic liquid, R, will be consumed by the reaction prior to the forsterite, and the resultant mixture will consist of forsterite and enstatite. However, in the case in which the starting composition is Y, which is enriched in silica relative to enstatite, the forsterite will be depleted before the liquid, and the reaction will yield the liquid and enstatite. Only in the case where the starting composition matches that of enstatite will the liquid and the forsterite be consumed at the same time, leaving only enstatite. The starting composition X represents the most common crystallization behaviour for saturated tholeiitic basaltic magmas; consequently, these magmas will experience a reaction between the liquid and the olivine, forsterite, at some point during their crystallization. This means that the liquid will be consumed by the reaction with forsterite and crystallization will cease. If, however, forsterite can be removed physically from the liquid before the reaction can occur, the reaction will be prevented and the peritectic liquid will remain to crystallize the pyroxene, enstatite, and move down toward the eutectic temperature where cristobalite and enstatite will crystallize.

The Albite-anorthite System

Most of the common minerals found in igneous rocks are solid-solution phases. These include olivine, pyroxene, amphibole, biotite, and plagioclase feldspars. Crystallization behaviour is illustrated best by using the $NaAlSi3O8$ (albite or Ab)–$CaAl_2Si_2O_8$ (anorthite or A_n) plagioclase system. Consider a liquid of composition L (60 percent A_n + 40 percent Ab) which is at an initial temperature of 1,500 °C. On cooling it will begin crystallizing plagioclase with 85 percent An (point P on the solidus) at the liquidus temperature of about 1,470 °C. As cooling continues further, the liquid will move down the liquidus toward B while simultaneously reacting continuously with the early-formed plagioclase to convert it to a homogeneous plagioclase that is more albitic and in equilibrium with the liquid. For example, when the liquid has reached A, at 1,400 °C, about 65 percent plagioclase with about 73 percent A_n (point O on the solidus) has crystallized from the liquid, which is now at about 36 percent An and 64 percent Ab. Finally, when the temperature of about 1,330 °C is reached, the last small amount of the liquid of composition 20 percent A_n + 80 percent Ab is consumed in the reaction and a homogeneous plagioclase of 60 percent A_n + 40 percent Ab remains (point S). Now consider the case in which the liquid is prevented from reacting with the early-formed plagioclase. This may be achieved by physically removing the plagioclase immediately after its formation or by cooling the liquid faster than the reaction process can consume the plagioclase. The liquid could theoretically reach the pure A_b composition at 1,100 °C, where it will disappear into the crystallizing albite. A whole range of plagioclase compositions from A_{n84} to A_{n00} will be preserved in the cooling process.

Bowen's Reaction Series

These two examples illustrate two principal reactions that occur during crystallization of common magmas, one discontinuous (the olivine-liquid-pyroxene reaction) and the other continuous (the plagioclase-liquid reaction). This was recognized first by the American petrologist Norman L. Bowen, who arranged the reactions in the form shown in figure; in his honour, the mineral series has since been called the Bowen's reaction series. The left branch of the Y-shaped arrangement consists of the discontinuous series that begins with olivine at the highest temperature and progresses through pyroxene, amphibole, and biotite as the temperature decreases. This series is discontinuous because the reaction occurs at a fixed temperature at constant pressure wherein the early-formed mineral is converted to a more stable crystal. Each mineral in the series displays a different silicate structure that exhibits increased polymerization as the temperature drops; olivine belongs to the island silicate structure type; pyroxene, the chain; amphibole, the double chain; and biotite, the sheet. On the other hand, the right branch is the continuous reaction series in which plagioclase is continuously reacting with the liquid to form a more albitic phase as the temperature decreases. In both cases, the liquid is consumed in the reaction.

When the two reaction series converge at a low temperature, minerals that will not react with the remaining liquid approach eutectic crystallization. Potash feldspar, muscovite, and quartz are crystallized. The phases that are crystallized first are the common minerals that compose basalt or gabbro, like bytownite or labradorite with pyroxene and minor amounts of olivine. Andesite or diorite minerals, such as andesine with either pyroxene or amphibole, crystallize next and are followed by orthoclase and quartz, which are the essential constituents of rhyolite or granite. A basaltic liquid at the top of the Y can descend to the bottom of the series to crystallize quartz only if the earlier reactions are prevented. As demonstrated above, complete reactions between early-formed minerals and the liquid depletes the supply of the liquid, thereby curtailing the progression down the series. One means by which basaltic magma can be transformed to rocks lower in the series is by fractional crystallization. In this process, the early-formed minerals are removed from the liquid by gravity (such minerals as olivine and pyroxene are denser than the liquid from which they crystallized), and so unreacted liquid remains later in the series.

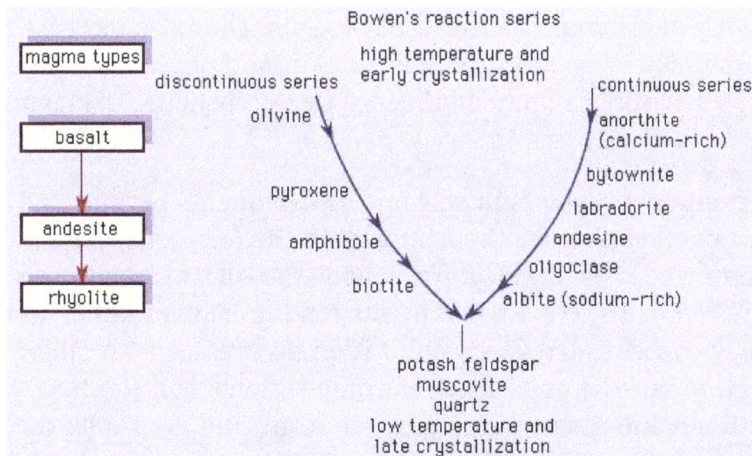

Bowen's reaction series showing the sequence of minerals that would be formed and removed during fractional crystallization of a melt. The magmas relating to the crystallizing minerals are shown on the left.

Assimilation

Another method of creating different daughter magmas from a parent is by having the latter react with its wall rocks. Consider a magma that is crystallizing pyroxene and labradorite. If the magma tears from its wall minerals, say, olivine and anorthite, which are formed earlier than pyroxene and labradorite in the series, they will react with the liquid to form these same minerals with which the magma is in equilibrium. The heat for driving this reaction comes directly from the magma itself. More pyroxene and labradorite will crystallize during the reaction and will release their latent heats of crystallization. On the other hand, if a mineral (quartz, for example) formed at a later stage than pyroxene or labradorite falls from the rock wall into the magma, the latent heat provided by further crystallization of pyroxene and labradorite will cause it to dissolve. This situation will occur only if the quartz from the wall rock is at a lower temperature than the magma. It will cause the magma to transfer its heat to the quartz in a cooling process. The cooling of the magma will necessarily be accompanied by the crystallization of the minerals already present. In both cases, the composition of the parent magma will be changed by the xenolithic (foreign rock) contamination. The contaminant need not belong to the reaction series in order for it to cause reactions or dissolution. In most cases, the end result will be a shift from the original composition of the parent magma toward that of the contaminant. This process in which wall rocks are incorporated into the magma is called assimilation. Because assimilation is accompanied by crystallization, it is likely that both fractional crystallization and assimilation will take place simultaneously. This combined process, referred to as AFC for assimilation–fractional crystallization, has been proposed as the mechanism by which andesites are produced from basalts.

Volatile Constituents and Late Magmatic Processes

Effects of Water and other Volatiles

Water and most other volatile substances profoundly influence the properties and behaviour of magmas in which they are dissolved. They reduce viscosity, lower temperatures of crystallization by tens to hundreds of degrees, and participate directly in the formation of minerals that contain essential hydroxyl (OH) or elements such as the halogens. They also increase rates of crystallization and reaction, especially when they are present as a fluid phase distinct from the magma. In general, however, they have only a limited influence on the sequence of magmatic crystallization, except in the latest stages of the reaction series.

The relatively low confining pressures in volcanic environments permit ready escape of volatile constituents, which nonetheless leave their imprint in the form of special mineral assemblages and a variety of textural and structural features among the volcanic rocks. Under the higher pressures of plutonic environments, these constituents tend to be maintained in magmatic solution and to be increasingly concentrated as crystallization progresses with falling temperature. Few members of the reaction series require them as compositional contributors; water, for example, is not thus used until amphiboles or micas begin to form, and even then the amounts removed from the melt rarely are large. Escape of volatiles from the system can occur "osmotically" if the enclosing rocks are pervious to them but not to the magma, but in general they are fractionated in favour of the residual melt until their concentration reaches the limit of solubility under the prevailing conditions of temperature and effective confining pressure. When this happens, normally at a very late stage of magmatic crystallization, they are exsolved from the melt as a separate fluid phase that under most circumstances is a supercritical gas. This process has been referred to as resurgent boiling, a somewhat misleading term because the exsolved fluid is not necessarily expelled from the system.

Pegmatites and Late-stage Mineralization

Coexistence of residual magma and a volatile-rich fluid (generally aqueous) promotes the partitioning and segregation of constituents, as well as the growth of very large crystals. The exsolved fluid, with its very low viscosity, not only can move readily through open spaces in the nearly solid igneous rock and in adjacent rocks but also serves as a medium through which various substances can diffuse rapidly in response to concentration gradients. Thus, it plays an important role in the formation of such special rock types as the pegmatites and lamprophyres, special features such as miaroles and plumose mineral aggregates, and many kinds of ore deposits whose constituents are derived from the original magma.

Most plutonic systems remain at elevated temperatures for long periods of time after all magma has been used up, and during these periods hydrothermal conditions normally obtain. These depend upon the continued presence of a typically aqueous fluid that further facilitates crystallization and exchanges of materials. It speeds up exsolution within homogeneous solid phases and devitrification of any glass that may be present, and it is a potent agent in the alteration, leaching, and replacement of minerals. Rock textures thereby are modified, especially along boundaries between original mineral grains, and details of composition also can be much changed. In some instances the bulk chemistry of the rock is markedly affected.

The hydrothermal alterations favour development of phases such as albite, carbonates, chlorites, clay minerals, epidotes, iron oxides, micas, silica minerals, talc, and zeolites, and many of them are accompanied by gross changes in volume.

Forms of Occurrence

Extrusive Igneous Rocks

Extrusive igneous rocks are the products of volcanic activity. They appear at the surface as molten lava that spreads in sheets and hardens, or they are made up of fragments of magma ejected from vents by violent gaseous explosions. Large-scale extrusive features include stratovolcanoes (composite cones), shield volcanoes, lava domes, and cinder cones. Smaller extrusive features include lava flows known as pahoehoe.

Intrusive Igneous Rocks

Erosion of volcanoes will immediately expose shallow intrusive bodies such as volcanic necks and diatremes. A volcanic neck is the "throat" of a volcano and consists of a pipelike conduit filled with hypabyssal rocks. Ship Rock in New Mexico and Devil's Tower in Wyoming are remnants of volcanic necks, which were exposed after the surrounding sedimentary rocks were eroded away. Many craterlike depressions may be filled with angular fragments of country rock (breccia) and juvenile pyroclastic debris. When eroded, such a depression exposes a vertical funnel-shaped pipe that resembles a volcanic neck with the exception of the brecciated filling. These pipes are dubbed diatremes. Many diatremes are formed by explosion resulting from the rapid expansion of gas—carbon dioxide and water vapour. These gases are released by the rising magma owing to the decrease in pressure as it nears the surface. Some diatremes contain kimberlite, a peridotite that contains a hydrous mineral called phlogopite. Kimberlite may contain diamonds.

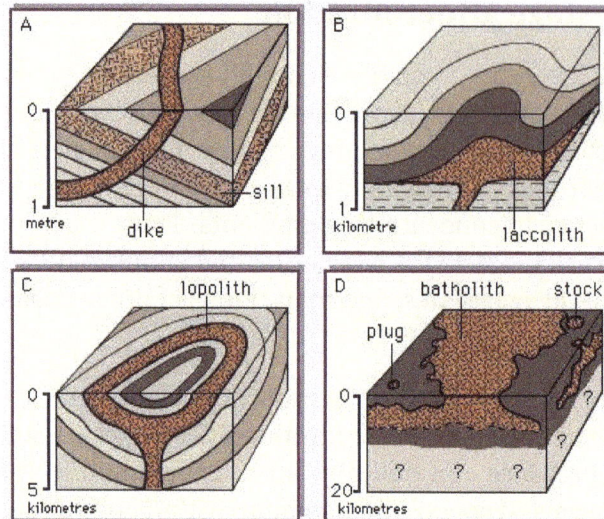

Forms of intrusive igneous rock bodies in hypothetical sections of Earth strata.

Dikes are usually tabular bodies that may radiate from the central vent of a volcano or from a volcanic neck. Not all dikes are associated with volcanoes, but they can be distinguished by their discordant relationship with the structure of the country rock that they cut across. Many dikes are only a few metres wide, but large ones, such as the dike that feeds the Muskox intrusion in the Northwest Territories of Canada, reach widths of more than 150 metres. Related to dikes are features that maintain a concordant relationship with the structure of the country rocks. Magmas may force their way between layers of country rock and solidify parallel to them to form sills. On the west bank of the Hudson River opposite New York City, the 300-metre-thick Palisades sill is exposed and can be traced for 80 kilometres. A laccolith also is concordant with country rock, but it is distinguished from a sill by having a flat floor with a domed (mushroom-shaped) roof. Laccoliths were first described in the Henry Mountains of Utah, where they may measure up to 200 metres thick with basal diameters exceeding three kilometres. Rocks of intermediate silica content generally make up these domed intrusions. In contrast, lopoliths are saucer-shaped bodies with a concave upward roof and floor and are commonly composed of mafic rocks. Lopoliths are huge in size; the Bushveld intrusive complex in South Africa, for example, has an area of about 66,000 square kilometres and an exposed thickness of 8 kilometres. The Muskox intrusion, mentioned above, is another large lopolith, which is estimated to be about 80 kilometres long and 11 kilometres wide (roof rocks covering part of the intrusion prevent an exact measurement).

These lopoliths are commonly layered with igneous minerals and rocks; in the Bushveld intrusion, one layer about 1 metre thick consisting of almost pure chromite (an ore of chromium) extends for tens of kilometres. Large irregularly shaped plutons are called either stocks or batholiths, depending on their sizes. Plutons larger than 100 square kilometres in area are termed batholiths, while those of lesser size are called stocks. It may be possible, however, that some stocks are the visible portions of batholiths that have not been exposed by erosion. Batholiths (from the Greek word bathos, meaning depth) are deep-seated crustal intrusions, whereas stocks may be formed at shallow depths only a few kilometres below the surface. Rocks ranging from quartz diorite to granite are commonly found in batholiths. Large batholiths in North America include the Sierra Nevada, the Idaho, and the Coast Range, which is about 600 kilometres long and 200 kilometres wide and extends from the Alaskan border through British Columbia to Washington state. Many pulses of intrusions contribute to the formation of these large bodies; for example, eight episodes of activity have been recognized in the Sierra Nevada batholith. They are formed, therefore, by the coalescence of many smaller batholiths and stocks.

Distribution of Igneous Rocks on Earth'S Surface

Divergent Plate Boundaries

Most of the igneous activity on Earth is restricted to a narrow zone that is related intimately with the motions of the lithospheric plates. Indeed, the composition of the magma, the types of volcanism, and the characteristics of intrusions are governed to a large extent by plate tectonics. The magmatism at divergent plate boundaries along the crests of the oceanic rises and ridges is mostly unseen except in places where the volcanic activity occurs subaerially (e.g., Iceland, which sits on the Mid-Atlantic Ridge). Along these divergent boundaries, the erupted basalts have such a restricted compositional range that they are referred to as mid-ocean-ridge basalt (MORB). They are subalkaline tholeiites that contain olivine in the norm and less than 0.25 percent potash. The chemistry suggests that MORB was generated from a mantle that was depleted of volatile elements (e.g., lanthanum [La], cerium [Ce], sodium, and potassium) in a previous partial melting process. A wide rift valley marks the crest of most of the oceanic ridges and rises. The valley is bounded by faults created by the divergent forces and is floored in its centre by a fracture zone (a mass of rock with many small breakages). These faults and fractures are the conduits for the MORB magmas that flood the valley, build volcanoes, and produce dikes by filling the conduits. Layer 2 of the oceanic crust results from these magmatic activities. As the plates diverge, MORB becomes the ocean floor on which oceanic sediments (layer 1) are deposited. This makes MORB the most abundant rock on the surface of Earth.

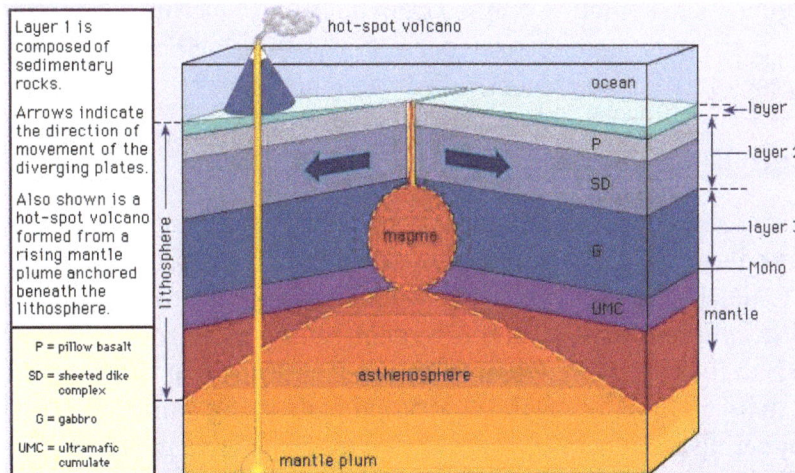

Idealized cross section of a divergent plate boundary showing the structure of the oceanic lithosphere.

Below the collection of lavas and dikes in layer 2 are found gabbro and diorite. They represent the plutonic rocks formed as a result of differentiation of the MORB magma that fed the volcanic activity along the rift. (Differentiation is the process in which more than one rock type is derived from a single parent magma.) These coarse-grained intrusives account for about 4 to 5 kilometres of layer 3, which rests on a sequence of layered ultramafic rocks. The rocks were formed by the gravitative accumulation of mafic minerals from the original MORB magma that filled a large chamber below the ridge axis. Below this layered sequence is mantle rock that is highly deformed and depleted (of elements such as lanthanum, cerium, sodium, and potassium that have been removed by repeated partial melting). Because seismic waves cannot distinguish between layered ultramafic rocks, which are not true mantle rocks, and ultramafic mantle rocks, the Moho actually is positioned between layer 3 and the layered ultramafics. The sequences consisting of layer 1 (limestone and chert sedimentary rocks), layer 2 of MORB lavas and dikes, and layer 3 of gabbro and diorite and the ultramafic rocks are known as ophiolites. Many geologists believe that ophiolites formed at oceanic ridges were emplaced by tectonic forces at convergent plate boundaries and then became exposed in highly deformed orogenic (mountain) belts. In fact, the same sequences of rocks were first reported in the Alps and were considered deep-seated intrusions. Some geologists still argue that all ophiolites were not formed at divergent plate boundaries.

Away from the axis of divergence, the composition of the volcanic rocks becomes more diverse. Most of the magmatism is related to hot spots, which are hot rising plumes of mantle rock that are anchored beneath the moving lithospheric plates. The Hawaiian Islands owe their existence to the magmatism associated with a hot spot that currently is located just southeast of the large island of Hawaii. This mantle plume not only provides magma for the eruptions at Kilauea Volcano but also is responsible for the submarine volcano named Loihi that will eventually become a new island. Most of the islands are built on a tholeiite basalt base, but the caps of the volcanoes are alkali basalts. The final episodes of volcanic activity on an island are extremely undersaturated; nephelinites and olivine melilite nephelinites are common products. The alkali basalts have differentiated to more silica-rich compositions, with hawaiites, mugearites, and trachytes being erupted in minor amounts. The two active volcanoes on Hawaii, Mauna Loa and Kilauea, are still erupting tholeiite basalts. Tholeiites on all the islands far from the ocean ridge crests are different from MORB in that they are enriched in lanthanum, cerium, sodium, and potassium. Early in Earth's history, a high-magnesium, high-temperature mafic magma called komatiite erupted from hot spots. Since most komatiites are only found in Archean regions, they are thought to be evidence for Earth being hotter than when it was initially formed. The youngest komatiite was recently discovered on the island of Gorgona, Colom.

Convergent Plate Boundaries

Igneous rocks associated with convergent plate boundaries have the greatest diversity. In this case, granite batholiths underlie the great composite volcanoes and consist of rocks ranging from basalt through andesite to dacite and rhyolite. These boundaries are destructive and consume the subducting oceanic lithosphere formed at the divergent centres. The rocks generated, however, are added on (accreted) to the continent. Oceanic trenches outline the junction of the colliding plates, but the igneous activity takes place on the overriding plate along a line at least about 100 kilometres above the subducting plate. In other words, almost no volcanism occurs between this 100-kilometre line (called the volcanic front) and the trench. The horizontal distance between the trench and the volcanic front depends on the angle of subduction; the steeper the angle, the shorter the distance. Volcanism occurs from this volcanic axis inland for a few hundred kilometres. The dominant rock constituting the composite volcanoes is andesite, but in some younger island arcs basalt tends to be more common, and in older volcanic areas dacite or rhyolite becomes prominent. Two different series of rocks are found in some volcanic chains. In Japan a tholeiitic series and a calc-alkalic series sometimes erupt from the same volcano. The former is characterized by lower magnesium, potassium, nickel, chromium, uranium, and thorium and a higher iron:magnesium ratio. Mineralogically, the tholeiitic series characteristically contains pigeonite (a low-calcium monoclinic pyroxene) in the groundmass of the basalts and andesites. The calc-alkalic series lacks pigeonite but instead has hypersthene. Most of the composite volcanoes of the Cascades Range in Oregon and Washington in the northwestern United States are characteristically calc-alkalic. In some volcanic arcs in areas farthest from the trench, a potassic series is found. In Japan the volcanoes within the Sea of Japan and farthest from the Japan Trench have alkali basalt compositions. Recent discoveries in modern convergent margins have identified igneous rocks within the oceanic trench sediments. These occur in regions where a mid-ocean ridge is being subducted. This creates higher heat flow and different types of igneous rocks, termed trondhjemite-tonalite-dacite (TTD) suites and alkaline, mafic, and felsic types.

In older areas of convergence, the composite volcanoes have been eroded, exposing the deeper plutonic granite batholiths that extend the entire length of the convergent boundaries. The batholiths are predominantly granodiorite, but gabbro through granite occur as well. It seems anomalous to find diorite, the plutonic equivalent of andesite, in low abundance since andesite is the dominant rock type of the volcanoes that were above these batholiths. Two basic types of granite have been recognized. The more common variety is located closer to the trench, has hornblende as its mafic mineral, is enriched in sodium and calcium, and has mantle chemical signatures; it is called I-type granite. The other type, called S-type granite, has muscovite and biotite and is depleted in sodium but enriched in aluminum such that corundum occurs in the norm and isotopic signatures. This suggests that such granites were formed by partial fusion of sedimentary rocks.

Flood Basalts

On the continental plates at areas away from active convergence, the magmatism is confined to rift valleys and local hot spots. The volume of magma produced is minor in comparison to that generated at oceanic rises and at convergent plate boundaries. Flood basalts are the most common form of occurrence. They span the rock record from the Precambrian to the Neogene Period (from about 4.6 billion to 2.6 million years ago) and are found worldwide. The 1.1-billion-year-old Keweenawan flood basalts in the Lake Superior region of northern Michigan may have formed in a rift that failed. The rifting of Pangaea that began during Jurassic time (approximately 200 million to 146 million years ago) generated flood basalt eruptions all along the newly opened Atlantic Ocean. Two voluminous eruptions associated with the opening of the South Atlantic produced the Paraná basalt in Brazil and the Karoo (or Karroo) in South Africa. The Deccan basalts in India were formed in the rift valleys associated with the breakup of Gondwana during the Cretaceous Period (approximately 146 million to 65.5 million years ago). Chemically, the most abundant basalts are supersaturated tholeiites with normative quartz, but olivine tholeiites and alkali basalts also are found. Feeder dike swarms (groups consisting of many parallel dikes) and sills are common in flood basalt plateaus. Alkaline rocks, such as those found in the East African Rift System, occur as well but are less abundant. This rift system stretches southward from the Red Sea–Gulf of Aden to Lake Victoria. Undersaturated basalts are most common in these rifts. During one eruption, a magma composed mostly of sodium carbonate issued from a volcanic vent that.

Sedimentary Rocks

Sedimentary rocks that are formed at or near the Earth's surface by the accumulation and lithification of sediment (detrital rock) or by the precipitation from solution at normal surface temperatures (chemical rock). Sedimentary rocks are the most common rocks exposed on the Earth's surface but are only a minor constituent of the entire crust, which is dominated by igneous and metamorphic rocks.

Sedimentary rocks are produced by the weathering of preexisting rocks and the subsequent transportation and deposition of the weathering products. Weathering refers to the various processes of physical disintegration and chemical decomposition that occur when rocks at the Earth's surface are exposed to the atmosphere (mainly in the form of rainfall) and the hydrosphere. These processes produce soil, unconsolidated rock detritus, and components dissolved in groundwater and runoff. Erosion is the process by which weathering products are transported away from the weathering site, either as solid material or as dissolved components, eventually to be deposited as sediment. Any unconsolidated deposit of solid weathered material constitutes sediment. It can form as the result of deposition of grains from moving bodies of water or wind, from the melting of glacial ice, and from the downslope slumping (sliding) of rock and soil masses in response to gravity, as well as by precipitation of the dissolved products of weathering under the conditions of low temperature and pressure that prevail at or near the surface of the Earth.

Sedimentary rocks are the lithified equivalents of sediments. They typically are produced by cementing, compacting, and otherwise solidifying preexisting unconsolidated sediments. Some varieties of sedimentary rock, however, are precipitated directly into their solid sedimentary form and exhibit no intervening existence as sediment. Organic reefs and bedded evaporites are examples of such rocks. Because the processes of physical (mechanical) weathering and chemical weathering are significantly different, they generate markedly distinct products and two fundamentally different kinds of sediment and sedimentary rock: (1) terrigenous clastic sedimentary rocks and (2) allochemical and orthochemical sedimentary rocks.

Clastic terrigenous sedimentary rocks consist of rock and mineral grains, or clasts, of varying size, ranging from clay-, silt-, and sand- up to pebble-, cobble-, and boulder-size materials. These clasts are transported by gravity, mudflows, running water, glaciers, and wind and eventually are deposited in various settings (e.g., in desert dunes, on alluvial fans, across continental shelves, and in river deltas). Because the agents of transportation commonly sort out discrete particles by clast size, terrigenous clastic sedimentary rocks are further subdivided on the basis of average clast diameter. Coarse pebbles, cobbles, and boulder-size gravels lithify to form conglomerate and breccia; sand becomes sandstone; and silt and clay form siltstone, claystone, mudrock, and shale.

Chemical sedimentary rocks form by chemical and organic reprecipitation of the dissolved products of chemical weathering that are removed from the weathering site. Allochemical sedimentary rocks, such as many limestones and cherts, consist of solid precipitated nondetrital fragments (allochems) that undergo a brief history of transport and abrasion prior to deposition as nonterrigenous clasts. Examples are calcareous or siliceous shell fragments and oöids, which are concentrically layered spherical grains of calcium carbonate. Orthochemical sedimentary rocks, on the other hand, consist of dissolved constituents that are directly precipitated as solid sedimentary rock and thus do not undergo transportation. Orthochemical sedimentary rocks include some limestones, bedded evaporite deposits of halite, gypsum, and anhydrite, and banded iron formations.

Sediments and sedimentary rocks are confined to the Earth's crust, which is the thin, light outer solid skin of the Earth ranging in thickness from 40–100 kilometres (25 to 62 miles) in the continental blocks to 4–10 kilometres in the ocean basins. Igneous and metamorphic rocks constitute the bulk of the crust. The total volume of sediment and sedimentary rocks can be either directly measured using exposed rock sequences, drill-hole data, and seismic profiles or indirectly estimated by comparing the chemistry of major sedimentary rock types to the overall chemistry of the crust from which they are weathered. Both methods indicate that the Earth's sediment-sedimentary rock shell forms only about 5 percent by volume of the terrestrial crust, which in turn accounts for less than 1 percent of the Earth's total volume. On the other hand, the area of outcrop and exposure of sediment and sedimentary rock comprises 75 percent of the land surface and well over 90 percent of the ocean basins and continental margins. In other words, 80–90 percent of the surface area of the Earth is mantled with sediment or sedimentary rocks rather than with igneous or metamorphic varieties. The sediment-sedimentary rock shell forms only a thin superficial layer. The mean shell thickness in continental areas is 1.8 kilometres; the sediment shell in the ocean basins is roughly 0.3 kilometre. Rearranging this shell as a globally encircling layer (and depending on the raw estimates incorporated into the model), the shell thickness would be roughly 1–3 kilometres.

Despite the relatively insignificant volume of the sedimentary rock shell, not only are most rocks exposed at the terrestrial surface of the sedimentary variety, but many of the significant events in Earth history are most accurately dated and documented by analyzing and interpreting the sedimentary rock record instead of the more voluminous igneous and metamorphic rock record. When properly understood and interpreted, sedimentary rocks provide information on ancient geography, termed paleogeography. A map of the distribution of sediments that formed in shallow oceans along alluvial fans bordering rising mountains or in deep, subsiding ocean trenches will indicate past relationships between seas and landmasses. An accurate interpretion of paleogeography and depositional settings allows conclusions to be made about the evolution of mountain systems, continental blocks, and ocean basins, as well as about the origin and evolution of the atmosphere and hydrosphere. Sedimentary rocks contain the fossil record of ancient life-forms that enables the documentation of the evolutionary advancement from simple to complex organisms in the plant and animal kingdoms. Also, the study of the various folds or bends and breaks or faults in the strata of sedimentary rocks permits the structural geology or history of deformation to be ascertained.

Finally, it is appropriate to underscore the economic importance of sedimentary rocks. For example, they contain essentially the world's entire store of oil and natural gas, coal, phosphates, salt deposits, groundwater, and other natural resources.

Several subdisciplines of geology deal specifically with the analysis, interpretation, and origin of sediments and sedimentary rocks. Sedimentary petrology is the study of their occurrence, composition, texture, and other overall characteristics, while sedimentology emphasizes the processes by which sediments are transported and deposited. Sedimentary petrography involves the classification and study of sedimentary rocks using the petrographic microscope. Stratigraphy covers all aspects of sedimentary rocks, particularly from the perspective of their age and regional relationships as well as the correlation of sedimentary rocks in one region with sedimentary rock sequences elsewhere.

Classification Systems

In general, geologists have attempted to classify sedimentary rocks on a natural basis, but some schemes have genetic implications (i.e.,knowledge of origin of a particular rock type is assumed), and many classifications reflect the philosophy, training, and experience of those who propound them. No scheme has found universal acceptance, and discussion here will centre on some proposals.

The book Rocks and Rock Minerals by Louis V. Pirsson was first published in 1908, and it has enjoyed various revisions. Sedimentary rocks are classified there rather simplistically according to physical characteristics and composition into detrital and nondetrital rocks.

Numerous other attempts have been made to classify sedimentary rocks. The most significant advance occurred in 1948 with the publication in the Journal of Geology of three definitive articles by the American geologists Francis J. Pettijohn, Robert R. Shrock, and Paul D. Krynine. Their classifications provide the basis for all modern discussion of the subject. The nomenclature associated with several schemes of classifying clastic and nonclastic rocks will be discussed in the following sections, but a rough division of sedimentary rocks based on chemical composition is shown in figure.

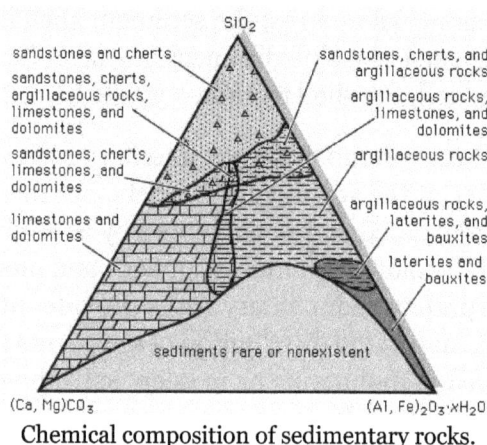

Chemical composition of sedimentary rocks.

For the purposes of the present discussion, three major categories of sedimentary rocks are recognized: (1) terrigenous clastic sedimentary rocks, (2) carbonates (limestone and dolomite), and (3) noncarbonate chemical sedimentary rocks. Terrigenous clastic sedimentary rocks are composed of the detrital fragments of preexisting rocks and minerals and are conventionally considered to be equivalent to clastic sedimentary rocks in general. Because most of the clasts are rich in silica, they are also referred to as siliciclastic sedimentary rocks. Siliciclastics are further subdivided on the basis of clast diameter into conglomerate and breccia, sandstone, siltstone, and finer-than-silt-sized mudrock (shale, claystone, and mudstone). The carbonates, limestones and dolomites, consist of the minerals aragonite, calcite, and dolomite. They are chemical sedimentary rocks in the sense that they possess at least in part a crystalline, interlocking mosaic of precipitated carbonate mineral grains. However, because individual grains such as fossil shell fragments exist for some period of time as sedimentary clasts, similar to transported quartz or feldspar clasts, most carbonates bear some textural affinities to the terrigenous clastic sedimentary rocks. The noncarbonate chemical sedimentary rocks include several rock types that are uncommon in the sedimentary rock record but remain important either from an economic point of view or because their deposition requires unusual settings. Specific varieties discussed below include siliceous rocks (cherts), phosphate rocks (phosphorites), evaporites, iron-rich sedimentary rocks (iron formations and ironstones), and organic-rich (carbonaceous) deposits in sedimentary rocks (coal, oil shale, and petroleum).

Despite the diversity of sedimentary rocks, direct measurement of the relative abundance of the specific types based on the study of exposed sequences suggests that only three varieties account for the bulk of all sedimentary rocks: mudrock, 47 percent; sandstone, 31 percent; and carbonate, 22 percent. Another method, which involves comparing the chemical composition of major sedimentary rock types with the chemistry of the Earth's continental crust, yields somewhat different numbers: mudrock, 79 percent; sandstone, 13 percent; and carbonate, 8 percent. Most sedimentary petrologists concede that the sedimentary rock record preserved and exposed within the continental blocks is selectively biased in favour of shallow-water carbonates and sandstones. Mudrocks are preferentially transported to the ocean basins. Consequently, indirect estimates based on chemical arguments are probably more accurate.

Terrigenous Clastic Rocks

A prominent physical feature of terrigenous clastic rocks is texture—that is, the size, shape, and arrangement of the constituent grains. These rocks have a fragmental texture: discrete grains are in tangential contact with one another. Terrigenous clastic sedimentary rocks are further subdivided on the basis of the mean grain diameter that characterizes most fragments, using the generally accepted size limits. Granules, pebbles, cobbles, boulders, and blocks constitute the coarse clastic sediments; sand-size (arenaceous) clasts are considered medium clastic sediments; and fine clastics sediments consists of silt- and clay-size materials.

The simplest way of classifying coarse clastic sedimentary rocks is to name the rock and include a brief description of its particular characteristics. Conglomerates and breccias differ from one another only in clast angularity. The former consist of abraded, somewhat rounded, coarse clasts, whereas the latter contain angular, coarse clasts. Thus, a pebble conglomerate is a coarse clastic sedimentary rock whose discrete particles are rounded and range from 4 to 64 millimetres (0.2 to 2.5 inches) in diameter. A more precise description reveals the rock types of the mineral fragments that compose the conglomerate—for example, a granite-gneiss pebble conglomerate.

Sandstones have long intrigued geologists because they are well exposed, are abundant in the geologic record, and provide an enormous amount of information about depositional setting and origin. Many classification schemes have been developed for sandstones, only the most popular of which are reviewed below. Most schemes emphasize the relative abundance of sand-size quartz, feldspar, and rock fragment components, as well as the nature of the material housed between this sand-size "framework" fraction.

Fine clastics are commonly, but rather simplistically, referred to as mudrocks. Mudrocks actually can include any clastic sedimentary rock in which the bulk of the clasts have diameters finer than 1/16 millimetre. Varieties include siltstone (average grain size between 1/16 and 1/256 millimetre) and claystone (discrete particles are mostly finer than 1/256 millimetre). Mud is a mixture of silt- and clay-size material, and mudrock is its indurated product. Shale is any fine clastic sedimentary rock that exhibits fissility, which is the ability to break into thin slabs along narrowly spaced planes parallel to the layers of stratification. Despite the great abundance of the fine clastics, disagreement exists as to what classification schemes are most useful for them, and an understanding of their origin is hindered by analytical complexities

Carbonate Rocks: Limestones and Dolomites

Limestones and dolostones (dolomites) make up the bulk of the nonterrigenous sedimentary rocks. Limestones are for the most part primary carbonate rocks. They consist of 50 percent or more calcite and aragonite (both $CaCO_3$). Dolomites are mainly produced by the secondary alteration or replacement of limestones; i.e., the mineral dolomite [CaMg(CO$_3$)$_2$] replaces the calcite and aragonite minerals in limestones during diagenesis. A number of different classification schemes have been proposed for carbonates, and the many categories of limestones and dolomites in the geologic record represent a large variety of depositional settings.

Noncarbonate Chemical Sedimentary Rocks

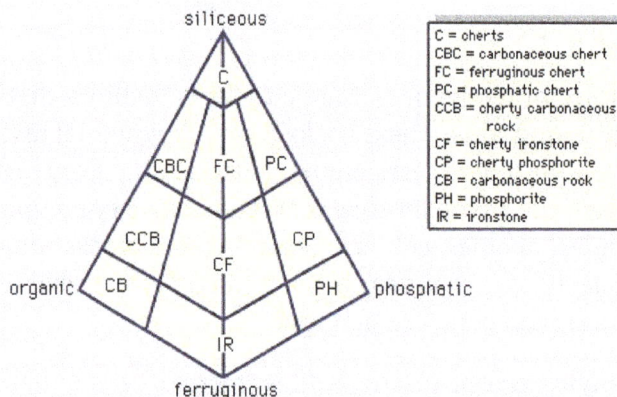

Common noncarbonate, nonclastic sedimentary rocks.

Noncarbonate chemical sedimentary rocks differ in many respects from carbonate sedimentary rocks and terrigenous clastic sedimentary rocks, and there is no single classification that has been universally accepted. This is a reflection of the great variation in mineral composition, texture, and other properties of these rock types. Such rocks as ironstones and banded iron formations (limonite, goethite, hematite, siderite, and chamosite), phosphorites, evaporites (rock salt, gypsum, and other salts), siliceous rocks (cherts), and organic-rich (carbonaceous) deposits of oil, natural gas, and coal in sedimentary rocks occur in much less abundance than carbonates and siliciclastic sedimentary rocks, although they may form thick and widespread deposits.

Classification schemes that incorporate all types of noncarbonate chemical sedimentary rocks do not exist because no triangular or tetrahedral scheme can accommodate all of them.

Properties of Sedimentary Rocks

Texture

Texture refers to the physical makeup of rock—namely, the size, shape, and arrangement (packing and orientation) of the discrete grains or particles of a sedimentary rock. Two main natural textural groupings exist for sedimentary rocks: clastic (or fragmental) and nonclastic (essentially crystalline). Noncarbonate chemical sedimentary rocks in large part exhibit crystalline texture, with individual mineral grains forming an interlocking arrangement. Depositional setting is an insignificant factor in both determining crystal size and altering crystalline texture. The size of crystals is controlled to a greater degree by the rate of precipitation, and their texture is modified by postdepositional recrystallization (reflecting the diagenetic environment). As a result, little attention is paid to crystalline textures other than providing a simple description of it (for example, coarsely crystalline versus finely crystalline). Also, even though carbonate rocks commonly include allochems that behave as clasts, they too are commonly diagenetically altered. Consequently, only cursory efforts are made to texturally characterize limestones and dolomites. Therefore, the following discussion deals in detail only with the textural techniques applied to terrigenous (siliciclastic) sedimentary rocks.

Grain Size

Particle size is an important textural parameter of clastic rocks because it supplies information on the conditions of transportation, sorting, and deposition of the sediment and provides some clues to the history of events that occurred at the depositional site prior to final induration. Determining the sizes of the discrete particles that constitute a sedimentary rock can be difficult, particularly if the rock is firmly indurated (cemented, compacted, and lithified). Various methods of measuring grain-size distribution have been devised; likewise several different grade-size schemes exist.

The size of particulate materials that make up sediments and sedimentary rocks are measured by weighing the proportions that accumulate in a series of wire mesh screen sieves, by visually counting grains with a petrographic microscope, or by determining the rate at which particles of varying diameter accumulate in a water-filled glass cylinder (known as a settling tube).

The standard ones used for sediments and sedimentary rocks. In the millimetre scale, each size grade differs from its predecessor by the constant ratio of 1:2; each size class has a specific class name used to refer to the particles included within it. This millimetre, or Udden-Wentworth, scale is a geometric grain-size scale since there is a constant ratio between class limits. Such a scheme is well suited for the description of sediments because it gives equal significance to size ratios, whether they relate to gravel, sand, silt, or clay. The phi scale is a useful, logarithmic-based modification of the Udden-Wentworth scale. Grain-size diameters in millimetres are converted to phi units using the conversion formula: phi $(\phi) = -\log_2 S$, where ϕ is size expressed in phi units and S is the grain size in millimetres. Phi values for grains coarser than one millimetre are negative, while those for grains finer than one millimetre are positive.

After the grain-size distribution for a given sediment or sedimentary rock has been determined by sieving, microscopic analysis, or use of a settling tube, it can be characterized using standard statistical measures in either of two ways: (1) visual inspection of various types of graphs that plot overall percent abundance versus grain-size diameter (e.g., histograms or bar diagrams, size frequency and cumulative size frequency curves, and probability curves that compare the actual grain-size distribution to a normal straight-line Gaussian distribution) or (2) arithmetic calculations made using diameter values in either millimetres or phi units that are read off the graphic plots and inserted into standard formulas.

For siliciclastic sedimentary rocks, the following standard statistical measures are conventionally described for grain-size distributions: (1) mode, the most frequently occurring particle size or size class, (2) median, the midpoint size of any grain-size distribution, (3) mean, an estimate of the arithmetic average particle size, (4) sorting or standard deviation, a measure of the range, scatter, or variation in grain size, (5) skewness, the degree of symmetry or asymmetry of the grain-size distribution, which is in turn a function of the coincidence or noncoincidence of mean, median, and mode, and (6) kurtosis (peakedness) of a grain-size distribution, which compares sorting in the central portion of the population with that in the tails.

Analysis of grain-size distribution is conducted with the disputed assumption that particular transporting agents and depositional settings (e.g., river delta deposits versus shallow marine longshore-bar sands) impose a distinctive textural "fingerprint" on the sediments they produce. Despite continuing efforts, the success of the various graphic and arithmetic approaches in characterizing grain-size distributions is debatable, as is their reliability in pinpointing ancient depositional settings. The grain-size distribution of sediments in many settings commonly appears to be inherited or to exhibit as much variation within a single environment as between different ones.

Particle Shape

Three different but related properties determine particle shape: form, roundness, and surface texture. Particle form is the overall shape of particles, typically defined in terms of the relative lengths of the longest, shortest, and intermediate axes. Particles can be spherical, prismatic, or bladelike. Roundness or angularity is a measure of the smoothness of particles. Surface texture refers to the presence or absence of small, variously shaped markings (pits, polish, scratches) that may occur on grain surfaces.

Each of these attributes of particle shape is traditionally measured in a standard fashion for the purpose of identifying the transporting agent and the depositional environment. Form is determined either by painstakingly measuring individual particles in three dimensions or by Fourier shape analysis, which uses harmonics analysis and computer digitizing to provide a precise description of particles in two dimensions. Form alone has limited usefulness in inferring depositional setting but more accurately reflects the mineralogy of the grains involved. Roundness is characterized by visually comparing grains to standard silhouette profiles. It is largely the result of abrasion history, which is controlled by the depositional agent and environment. For example, windblown and surf zone sands are well-rounded, while glacial sands and turbidity current deposits are angular. Particle roundness or angularity also reflects mineralogy (soft minerals are abraded more readily than hard minerals), clast size (coarse particles become rounded more rapidly than do fine ones), and transport distance (sands become more abraded and hence rounder as the distance traveled increases). Particle surfaces can be visually examined for pitting, markings, and polish through the use of a microscope or hand lens, or in some cases, a scanning electron microscope (SEM). Certain surface textures have been genetically linked to specific depositional agents; for example, classic V-shaped percussion marks identify quartz grains of the beach and nearshore zones.

Fabric

The fabric of a sedimentary rock controls the rock's porosity and permeability and therefore its ability to hold and transmit fluids such as oil and water. The orientation, or lack thereof, of the crystals or grains that make up a sedimentary rock constitutes one aspect of fabric. Genetically, there are two principal varieties of oriented fabrics: primary (or depositional) and secondary (or deformational). Primary fabrics are produced while the sediment is accumulating. For example, river currents and some submarine gravity flows generate sediments whose flaky and prismatic constituent particles have long or short axes parallel with one another to produce an oriented fabric. Secondary fabrics result from a rotation of the constituent elements under stress or from the growth of new elements during diagenesis. Fabrics in coarse clastic sedimentary rocks like conglomerates and sandstones can be determined by measuring and plotting dimensional directions, such as the long axes of pebbles or sand grains. In mudrocks, fabrics can be ascertained by studying the platelike arrangement of mica and clay minerals.

In addition to orientation, a factor known as packing contributes to a rock's fabric. Packing refers to the distribution of grains and intergranular spaces (either empty or filled with cement or fine-grained matrix) in a sedimentary rock. It is controlled by grain size and shape and by the degree of compaction of a sedimentary rock; in turn it determines the rock's bulk density. A description of packing is generally based on the analysis of thin sections of a sedimentary rock using a petro-graphic microscope. Particular attention is paid to the number of grain-to-grain contacts (packing proximity) and to comparisons between the sum of the lengths of grains to the total length of a traverse across a thin section (packing density).

Mineralogical and Geochemical Composition

Minerals that make up sedimentary rocks are of two principal types—namely, detrital and authi-genic. Detrital minerals, such as grains of quartz and feldspar, survive weathering and are trans-ported to the depositional site as clasts. Authigenic minerals, like calcite, halite, and gypsum, form in situ within the depositional site in response to geochemical processes. The chemical compounds that constitute them ultimately are generated by chemical weathering and are transported from the weathering site to the point of precipitation primarily in solution. Clay minerals are abundant in sedimentary rocks, particularly mudrocks, and some are detrital. They may have been produced at the weathering site by the partial decomposition of minerals like feldspar. They are transported as clasts, however, and thus can be regarded simply as fine- to very fine-textured detrital particles. Other clay minerals form authigenically at the site of deposition. Some of the important clay min-erals are kaolinite, halloysite, montmorillonite, illite, vermiculite, and chlorite.

The mean chemical composition of the major varieties of sedimentary rocks exhibits wide vari-ation Significant contrasts in overall composition among sandstones, carbonates, and mudrocks reflect fundamental differences not only in the mechanisms by which detrital minerals of different sizes are transported and deposited but also in the chemical conditions that permit precipitation of various authigenic minerals.

Diagenesis includes all physicochemical, biochemical, and physical processes (short of metamorphism) that modify sediments in the time between their deposition and their analysis. Lithification, the process by which sediment is converted into solid sedimentary rock, is one re-sult of diagenesis. Many diagenetic processes such as cementation, recrystallization, and dolomiti-zation are essentially geochemical processes; others like compaction are fundamentally physical processes. All diagenetic changes occur at the low temperatures and pressures characteristic of surface and near-surface environments. These changes can take place almost immediately after sediment formation, or they can occur hundreds or even millions of years later.

Sedimentary Structures

Sedimentary structures are the larger, generally three-dimensional physical features of sedimen-tary rocks; they are best seen in outcrop or in large hand specimens rather than through a micro-scope. Sedimentary structures include features like bedding, ripple marks, fossil tracks and trails, and mud cracks. They conventionally are subdivided into categories based on mode of genesis. Structures that are produced at the same time as the sedimentary rock in which they occur are called primary sedimentary structures. Examples include bedding or stratification, graded bedding, and cross-bedding. Sedimentary structures that are produced shortly after deposition and as a result of compaction and desiccation are called penecontemporaneous sedimentary structures. Examples include mud cracks and load casts. Still other sedimentary structures like concretions, vein fillings, and stylolites form well after deposition and penecontemporaneous modification; these are known as secondary structures. Finally, others like stromatolites and organic burrows and tracks, though they may in fact be primary, penecontemporaneous, or even secondary, may be grouped as a fourth category—organic sedimentary structures.

Considerable attention is paid to the sedimentary structures exhibited by any sedimentary rock. Primary sedimentary structures are particularly useful because their abundance and size suggest the probable transporting and depositional agents. Certain varieties of primary sedimentary structures like cross-bedding and ripple marks display orientations that are consistently related to the direction of current movement. Such structures are referred to as directional sedimentary structures because they can be used to infer the ancient paleocurrent pattern or dispersal system by which a sedimentary rock unit was deposited. Other sedimentary structures are stratigraphic "top and bottom" indicators. For example, the progressive upward decrease in clastic grain size diameters, known as graded bedding, would allow a geologist to determine which way is stratigraphically "up"—i.e., toward the younger beds in a dipping sedimentary bed. The suite (repeated sequence) of sedimentary structures in any single stratigraphic unit is another attribute by which that unit may be physically differentiated from others in the region.

External Stratification

Stratification (or bedding) is expressed by rock layers (units) of a general tabular or lenticular form that differ in rock type or other characteristics from the material with which they are interstratified (sometimes stated as interbedded, or interlayered). These beds, or strata, are of varying thickness and areal extent. The term stratum identifies a single bed, or unit, normally greater than one centimetre in thickness and visibly separable from superjacent (overlying) and subjacent (underlying) beds. "Strata" refers to two or more beds, and the term lamina is sometimes applied to a unit less than one centimetre in thickness. Thus, lamination consists of thin units in bedded, or layered, sequence in a natural rock succession, whereas stratification consists of bedded layers, or strata, in a geologic sequence of interleaved sedimentary rocks.

For most stratified sedimentary rocks, the arrangement of layers is one of unequal thickness, ranging from very thin laminae to discrete beds that measure a few to many metres in thickness. The terms thick and thin as applied to bedding, or stratification, are relative, reflecting the training of a particular geologist as well as experience with a specific stratigraphic section or sections.

Bedding Types and Bedding-plane Features

Types

It is common to discover a rhythmic pattern in a pile of stratified sedimentary rocks represented by a repetitive sequence of rock types. In most instances of such cyclic sedimentation, the bedding, or stratification, is horizontal or essentially so; that is, the transporting, sorting, and depositing agents of wind, running water, and lake and ocean currents and waves accumulated the laminae and strata in a flat-lying or horizontal arrangement. They are termed well-bedded, a type of primary stratification.

Primary stratification in sediments and sedimentary rocks can be cross-bedded (cross-stratified), graded, and imbricate and can also display climbing laminae, ripples, and beds.

Graded bedding simply identifies strata that grade upward from coarse-textured clastic sediment at their base to finer-textured materials at the top. The stratification may be sharply marked so that one layer is set off visibly from those above and beneath it. More commonly, however, the layers are blended. This variety of bedding results from a check in the velocity of the transporting agent, and thus coarse-textured sediment (gravel, for example) is deposited first, followed upward by pebbles, granules, sand, silt, and clay. It is commonly associated with submarine density currents.

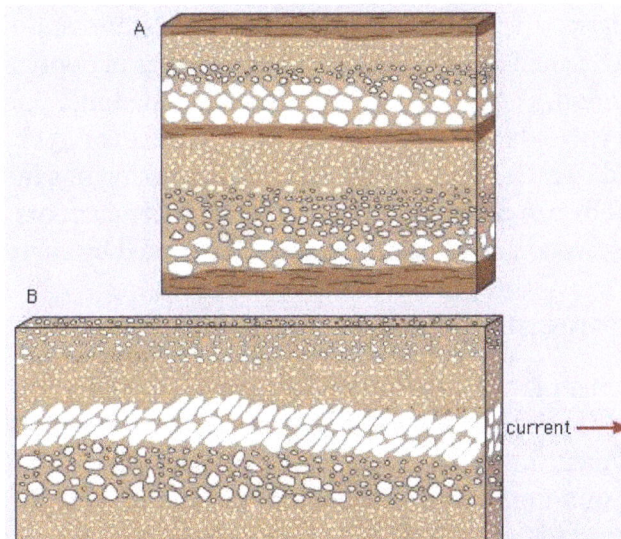

(A) Graded bedding. (B) Imbricate bedding.

Imbricate bedding is a shingle structure in a deposit of flattened or disk-shaped pebbles or cobbles. That is to say, elongated and commonly flattened pebbles and cobbles in gravelly sediment are deposited so that they overlap one another like roofing shingles. Imbricate bedding forms where high-velocity currents move over a streambed or where strong currents and waves break over a gradually sloping beach, thereby forming beach shingle.

Growth structures in sedimentary rocks are in situ features that accumulate largely as the result of organic buildups within otherwise horizontal or nearly flat-lying strata. Reefs and stromatolites are two common varieties of such growth structures.

Bedding-Plane Features

Upper surfaces of beds commonly display primary sedimentary features that are classified as bedding-plane structures. A three-dimensional view may be obtained if some of these can be seen from the side as well as from the top of a pile of strata. They include such features as ripples (ripple marks), climbing ripples, rills, pits, mud cracks, trails and tracks, salt and ice casts and molds, and others. Bedding-plane markings and irregularities can be allocated to one of three classes: (1) those on the base of a bed (load and current structures and organic markings), (2) those within a bed (parting lineation), and (3) those on top of a bed (ripple marks, pits, impressions, mud cracks, tracks and trails of organisms, and others).

Deformation Structures

In addition to sedimentary structures that are normally associated with bedding planes, there are other such structures that result from deformation during or shortly after sedimentation but before induration of the sediment into rock. These are nontectonic features—i.e., they are not bends and folds brought about by metamorphism or other such causes. Deformation structures can be grouped into several classes, as follows: (1) founder and load structures, (2) convoluted structures, (3) slump structures, (4) injection structures, such as sandstone dikes or sills, and (5) organic structures.

Structures found on the bottom of a bed are called sole markings, because they formed on the "sole" of the bed. Sole marks are commonly formed on sandstone and limestone beds that rest upon shale beds. They are termed casts, because they are fillings of depressions that formed on the surface of the underlying mud. They originate (1) by unequal loading upon the soft and plastic wet mud, (2) by the action of currents across the upper mud surface, or (3) by the activities of organisms on this surface. Load casts form as the result of downsinking of sandstone or limestone into the mud beneath. Current marks can form by the action of water currents on upper surfaces of the beds or by "tools" (such as wood and fossils) that are transported by currents over soft sediment.

Sedimentary Environments

The sedimentary environment is the specific depositional setting of a particular sedimentary rock and is unique in terms of physical, chemical, and biological characteristics. The physical features of a sedimentary environment include water depth and the velocity and persistence of currents. Chemical characteristics of an environment include the salinity (proportion of dissolved salts), acidity or basicity (pH), oxidation potential (Eh), pressure, and temperature. The biological characteristics are mainly the assemblage of fauna and flora that populate the setting. These conditions, combined with the nature of the transporting agent and the source area, largely determine the properties of the sediments deposited within the environment. A number of ways of classifying depositional environments exist, but most modern schemes employ a geomorphologic approach. That is to say, an environment is defined in terms of a distinct geomorphic unit or landform, modern examples of which are readily visible for comparative purposes—e.g., a river delta, an alluvial fan, a submarine fan, or the abyssal floor of an ocean basin.

Individual environments are further grouped into (1) marine environments, which include the nearshore, shallow littoral zone and the offshore, deep littoral zone, as well as deepwater realms, (2) mixed marine and nonmarine settings such as the beach and supratidal zones, and (3) nonmarine settings like lacustrine and various alluvial settings. Each environment is associated with a set of criteria that constitutes its distinguishing features.

Sedimentary Rock Types

Conglomerates and Breccias

Conglomerates and breccias are sedimentary rocks composed of coarse fragments of preexisting rocks held together either by cement or by a finer-grained clastic matrix. Both contain significant amounts (at least 10 percent) of coarser-than-sand-size clasts. Breccias are consolidated rubble;

their clasts are angular or subangular. Conglomerates are consolidated gravel whose clasts are subrounded to rounded. Sometimes the term rudite (or rudaceous) is used to collectively refer to both breccias and conglomerates.

Classification Schemes

A number of classification schemes have been proposed to further subdivide conglomerates and breccias. One scheme is purely descriptive, partitioning these coarse clastic sedimentary rocks on the basis of grain size (e.g., boulder breccia versus cobble conglomerate) or composition or both (chert pebble breccia versus limestone cobble conglomerate). Yet another scheme differentiates individual conglomerates and breccias according to depositional agency and environmental setting (alluvial fan conglomerate as opposed to beach conglomerate). The best classification systems incorporate objective physical characteristics of both composition and texture as well as mode of genesis. Conglomerates and breccias belong to four genetic categories: (1) epiclastic, produced by the physical disintegration (weathering) of preexisting rocks, (2) pyroclastic, produced by the explosive activity of volcanoes, (3) cataclastic, formed by local earth movements (fault breccias) or solution phenomena (collapse breccias), and (4) meteoritic, produced by the impact of extraterrestrial bodies on the Earth's surface. In a strict sense, epiclastic conglomerates and breccias are the only true sedimentary rocks, because they alone are produced by weathering.

Epiclastic Conglomerates and Breccias

There are two principal types of epiclastic conglomerates and breccias: intraformational, derived penecontemporaneously by eroding, transporting, and depositing material from within the depositional basin itself; and extraformational, derived from source rocks that lie outside the area in which the deposit occurs. Epiclastic conglomerates and breccias together probably make up no more than 1 or 2 percent of the conventional sedimentary rock record.

Intraformational conglomerates and breccias are widespread in the geologic record but are volumetrically unimportant. They occur as laterally continuous bands or horizons within sequences of shallow-water marine or nonmarine deposits. Their origin is commonly related to the existence of brief episodes of strong bottom-hugging currents capable of ripping up recently deposited, unconsolidated sediment. For example, shallow marine limestone deposits commonly have thin bands of boulder-, cobble-, and pebble-size carbonate clasts (edgewise conglomerate or breccia beds) that are generated when storm waves erode and redeposit carbonate mud layers. Likewise, high-velocity river currents that accompany torrential rains give rise to shale pebble conglomerates and breccias within sequences of floodplain alluvium. Other intraformational conglomerates and breccias mix shallow- and deep-water sedimentary rock clasts encased in a finer-grained matrix of deeper-water material. Such deposits accumulate as depositional aprons that flank the scarps (steep slopes) bounding shallow-water platformal areas such as the modern Great Bahama and Little Bahama banks off Florida in the southeastern United States.

Extraformational conglomerates and breccias, the coarse clastic sedimentary rocks derived by the weathering of preexisting rocks outside the depositional basin, are most important from the points of view of both volume and geologic significance. They can be subdivided into two specific categories: (1) clast-supported conglomerates (and breccias) and (2) matrix-supported conglomerates.

Clast-supported Conglomerates

These rocks contain less than 15 percent matrix—i.e., material composed of clasts finer than granule size (2-millimetre diameter or less). They typically exhibit an intact fabric that has a clast-supported framework such that the individual granules, pebbles, cobbles, and boulders touch each other. The space between framework components either is empty or is filled with chemical cement, finer-than-granule clasts, or both. Clast-supported conglomerates may be composed of large clasts of a single rock or mineral type (oligomictic orthoconglomerates), or they may contain a variety of rock and mineral clast types (petromictic orthoconglomerates). If the clasts are all of quartz, then it is called a quartzose conglomerate.

Clast-supported conglomerates (and orthobreccias) are deposited by highly turbulent water. For example, beach deposits commonly contain lenses and bands of oligomictic orthoconglomerate, composed mainly (95 percent or more) of stable, resistant, coarse clasts of vein quartz, quartzite, quartz sandstone, and chert. Such deposits are typically generated in the upper reaches of winter storm beaches where strong surf can sift, winnow, and abrade coarse pebbles and boulders. The most indestructible components are thereby consolidated as conglomeratic lenses that are interfingered with finer-grained, quartz-rich beach sandstones. Petromictic conglomerates and breccias, on the other hand, reflect the existence of high-relief (mountainous) source areas. Topographically high source areas signify tectonic mobility in the form of active folding or faulting or both. The existence of petromictic conglomerates and breccias in the geologic record is therefore significant: their presence and age not only pinpoint the timing and location of mountain building accompanied by sharp uplift and the possibility of regionally significant fault scarps but also can be used to infer the past distribution of physiographic features such as mountain fronts, continental block margins, continental shelf-continental slope boundaries, and the distribution of oceanic trenches and volcanic island arcs. Deep marine conglomerates may be called resedimented conglomerates. These were retransported from seashore areas by turbidity currents.

Bedding (layering and stratification) in clast-supported conglomerates, if apparent at all, is typically thick and lenticular. Graded bedding, in which size decreases from bottom to top, is common: because agitated waters rarely subside at once, declining transport power causes a gradual upward decrease in maximum clast size. Relative to the bedding, the pebbles in sandy conglomerates tend to lie flat, with their smallest dimension positioned vertically and the greatest aligned roughly parallel to the current. In closely packed orthoconglomerates, however, there is often a distinct imbrication; i.e., flat pebbles overlap in the same direction like roof shingles. Imbrication is upstream on riverbeds and seaward on beaches.

Clast-supported conglomerates are quite important economically because they hold enormous water reserves that are easily released through wells. This feature is attributable to their high porosity and permeability. Porosity is the volume percentage of "void" (actually fluid- or air-filled) space in a rock, whereas permeability is defined as the rate of flow of water at a given pressure gradient through a unit volume. The interconnectedness of voids in conglomerates contributes to their permeability. Also, because the chief resistance to flow is generally due to friction and capillary effects, the overall coarse grain size makes conglomerates even more permeable. The high degree of porosity and permeability causes conglomerates to generate excellent surface drainage, so they are to be avoided as dam and reservoir sites.

Matrix-supported Conglomerates

Matrix-supported conglomerates, also called diamictites, exhibit a disrupted, matrix-supported fabric; they contain 15 percent or more (sometimes as much as 80 percent) sand-size and finer clastic matrix. The coarse detrital clasts "float" in a finer-grained detrital matrix. They actually are mudrocks in which there is a sprinkling of granules, pebbles, cobbles, or boulders, or some combination of them. Accordingly, they are sometimes referred to as conglomerate mudstones or pebbly mudstones.

Although matrix-supported conglomerates originate in a variety of ways, they are not deposited by normal currents of moving water. Some are produced by submarine landslides, massive slumping, or dense, sediment-laden, gravity-driven turbidity flows. Matrix-supported conglomerates that can be definitively related to such mechanisms are called tilloids. Tilloids commonly make up olistostromes, which are large masses of coarse blocks chaotically mixed within a muddy matrix. The terms till (when unconsolidated) and tillite (when lithified) are used for diamictites that appear to have been directly deposited by moving sheets of glacial ice. Tillites typically consist of poorly sorted angular and subangular, polished and striated blocks of rock floating in an unstratified clay matrix. The clasts may exhibit a weak but distinct alignment of their long axes approximately parallel to the direction of ice flow. Tillites are notoriously heterogeneous in composition: clasts appear to be randomly mixed together without respect to size or compositional stability. These clasts are derived mainly from the underlying bedrock. Extremely coarse, far-traveled blocks and boulders are called erratics.

Other rarer diamictites, known as laminated pebbly (or cobbly or bouldery) mudstones, consist of delicately laminated mudrocks in which scattered coarser clasts occur. Laminations within the muddy component are broken and bent. They are located beneath and adjacent to the larger clasts but gently overlap or arch over them, suggesting that the coarse clasts are dropstones (i.e., ice-rafted blocks released as floating masses of ice melt).

Sandstones

Sandstones are siliciclastic sedimentary rocks that consist mainly of sand-size grains (clast diameters from 2 to 1/16 millimetre) either bonded together by interstitial chemical cement or lithified into a cohesive rock by the compaction of the sand-size framework component together with any interstitial primary (detrital) and secondary (authigenic) finer-grained matrix component. They grade, on the one hand, into the coarser-grained siliciclastic conglomerates and breccias described above, and, on the other hand, into siltstones and the various finer-grained mudrocks described below. Like their coarser analogues—namely, conglomerates and breccias—sand-size (also called arenaceous) sedimentary rocks are not exclusively generated by the physical disintegration of preexisting rocks. Varieties of limestone that contain abundant sand-size allochems like oöids and fossil fragments are, in at least a textural sense, types of sandstones, although they are not terrigenous siliciclastic rocks. Such rocks, called micrites when lithified or carbonate sands when unconsolidated, are more properly discussed as limestones. Also, pyroclastic sandstones or tuffs formed by lithifying explosively produced volcanic ash deposits can be excluded from this discussion because their origin is unrelated to weathering.

Sandstones are significant for a variety of reasons. Volumetrically they constitute between 10 and 20 percent of the Earth's sedimentary rock record. They are resistant to erosion and therefore greatly influence the landscape. When they are folded, they create the backbone of mountain ranges like the Appalachians of eastern North America, the Carpathians of east-central Europe, the Pennines of northern England, and the Apennine Range of Italy; when flat-lying, they form broad plains and plateaus like the Colorado and Allegheny plateaus. Sandstones are economically important as major reservoirs for both petroleum and water, as building materials, and as valuable sources of metallic ores. Most significantly, they are the single most useful sedimentary rock type for deciphering Earth history. Sandstone mineralogy is the best indicator of sedimentary provenance: the nature of a sedimentary rock source area, its composition, relief, and location. Sandstone textures and sedimentary structures also are reliable indexes of the transportational agents and depositional setting.

Sandstone Components and Colour

There are three basic components of sandstones: (1) detrital grains, mainly transported, sand-size minerals such as quartz and feldspar, (2) a detrital matrix of clay or mud, which is absent in "clean" sandstones, and (3) a cement that is chemically precipitated in crystalline form from solution and that serves to fill up original pore spaces.

The colour of sandstone depends on its detrital grains and bonding material. An abundance of potassium feldspar often gives a pink colour; this is true of many feldspathic arenites, which are feldspar-rich sandstones. Fine-grained, dark-coloured rock fragments, such as pieces of slate, chert, or andesite, however, give a salt-and-pepper appearance to sandstone. Iron oxide cement imparts tones of yellow, orange, brown, or red, whereas calcite cement imparts a gray colour. A sandstone consisting almost wholly of quartz grains cemented by quartz may be glassy and white. A chloritic clay matrix results in a greenish black colour and extreme hardness; such rocks are wackes.

Formation of Sandstones Today

Sandstones occur in strata of all geologic ages. Much scientific understanding of the depositional environment of ancient sandstones comes from detailed study of sand bodies forming at the present time. One of the clues to origin is the overall shape of the entire sand deposit. Inland desert sands today cover vast areas as a uniform blanket; some ancient sandstones in beds a few hundred metres thick but 1,600 kilometres or more in lateral extent, such as the Nubian Sandstone of North Africa, of Mesozoic age (about 245 to 66.4 million years old), also may have formed as blankets of desert sand. Deposits from alluvial fans form thick, fault-bounded prisms. River sands today form shoestring-shaped bodies, tens of metres thick, a few hundred metres wide, up to 60 kilometres or more long, and usually oriented perpendicularly to the shoreline. In meandering back and forth, a river may construct a wide swath of sand deposits, mostly accumulating on meander-point bars. Beaches, coastal dunes, and barrier bars also form "shoestring" sands, but these are parallel to the shore. Deltaic sands show a fanlike pattern of radial, thick, finger-shaped sand bodies interbedded with muddy sediments. Submarine sand bodies are diverse, reflecting the complexities of underwater topography and currents. They may form great ribbons parallel with the current; huge submarine "dunes" or "sand waves" aligned perpendicularly to the current; or irregular shoals, bars, and sheets. Some sands are deposited in deep water by the action of density currents, which flow down submarine slopes by reason of their high sediment concentrations and, hence, are called turbidity currents. These characteristically form thin beds interbedded with shales; sandstone beds often are graded from coarse grains at the base to fine grains at the top of the bed and commonly have a clay matrix.

Bedding Structure

One of the most fruitful methods of deciphering the environment of deposition and direction of transport of ancient sandstones is detailed field study of the sedimentary structures.

Bedding in sandstones, expressed by layers of clays, micas, heavy minerals, pebbles, or fossils, may be tens of feet thick, but it can range downward to paper-thin laminations. Flagstone breaks in smooth, even layers a few centimetres thick and is used in paving. Thin, nearly horizontal lamination is characteristic of many ancient beach sandstones. Bedding surfaces of sandstones may be marked by ripples (almost always of subaqueous origin), by tracks and trails of organisms, and by elongated grains that are oriented by current flow (fossils, plant fragments, or even elongated sand grains). Sand-grain orientation tends to parallel direction of the current; river-channel trends in fluvial sediments, wave-backwash direction in beach sands, and wind direction in eolian sediments are examples of such orientation.

A great variety of markings, such as flutes and scour and fill grooves, can be found on the undersides of some sandstone beds. These markings are caused by swift currents during deposition; they are particularly abundant in sandstones deposited by turbidity currents.

Within the major beds, cross-bedding is common. This structure is developed by the migration of small ripples, sand waves, tidal-channel large-scale ripples, or dunes and consists of sets of beds that are inclined to the main horizontal bedding planes. Almost all sedimentary environments produce characteristic types of cross-beds; as one example, the lee faces of sand dunes (side not facing the wind) may bear cross-beds as much as 33 metres (108 feet) high and dipping 35°.

Some sand stones contain series of graded beds. The grains at the base of a graded bed are coarse and gradually become finer upward, at which point there is a sharp change to the coarse basal layer of the overlying bed. Among the many mechanisms that can cause these changes in grain size are turbidity currents, but in general they can be caused by any cyclically repeated waning current.

After the sand is deposited, it may slide downslope or subside into soft underlying clays. This shifting gives rise to contorted or slumped bedding on a scale of centimetres to tens of metres. Generally these are characteristic of unstable areas of rapid deposition.

Local cementation may result in concretions of calcite, pyrite, barite, and other minerals. These can range from sand crystals or barite roses to spheroidal or discoidal concretions tens of metres across.

The fossil content also is a useful guide to the depositional environment of sandstones. Desert sandstones usually lack fossils. River-channel and deltaic sandstones may contain fossil wood, plant fragments, fossil footprints, or vertebrate remains. Beach and shallow marine sands contain mollusks, arthropods, crinoids, and other marine creatures, though marine sandstones are much less fossiliferous than marine limestones. Deepwater sands are frequently devoid of skeletal fossils, although tracks and trails may be common. The fossils are not actually structures, of course, but the living organisms were able to produce them. Burrowing by organisms, for example, may cause small-scale structures, such as eyes and pods or tubules of sand.

Texture

The texture of a sandstone is the sum of such attributes as the clay matrix, the size and sorting of the detrital grains, and the roundness of these particles. To evaluate this property, a scale of textural maturity that involved four textural stages was devised in 1951. These stages are described as follows.

Immature sandstones contain a clay matrix, and the sand-size grains are usually angular and poorly sorted. This means that a wide range of sand sizes is present. Such sandstones are characteristic of environments in which sediment is dumped and is not thereafter worked upon by waves or currents. These environments include stagnant areas of sluggish currents such as lagoons or bay bottoms or undisturbed seafloor below the zone of wave or current action. Immature sands also form where sediments are rapidly deposited in subaerial environments, such as river floodplains, swamps, alluvial fans, or glacial margins. Submature sandstones are created by the removal of the clay matrix by current action. The sand grains are, however, still poorly sorted in these rocks. Submature sandstones are common as river-channel sands, tidal-channel sands, and shallow submarine sands swept by unidirectional currents. Mature sandstones are clay-free, and the sand grains are subangular, but they are well sorted—that is, of nearly uniform particle size. Typically, these sandstones form in environments of current reversal and continual washing, such as beaches. Supermature sandstones are those that are clay-free and well sorted and, in addition, in which the grains are well rounded. These sandstones probably formed primarily as desert dunes, where intense eolian abrasion over a very long period of time may wear sand grains to nearly spherical shapes.

The methodology used for detailed study of siliciclastic sedimentary rock textures, particularly grain-size distribution and grain shape (angularity and sphericity) has been described above. The information that results from textural analyses is especially useful in identifying sandstone depositional environments. Dune sands in all parts of the world, for example, tend to be fine-sand-size (clast diameters from 1/4 to 1/8 millimetre) because sand of that dimension is most easily moved by winds. Desert (eolian) sandstones also tend to be bimodal or polymodal—i.e., having two (or more) abundant grain-size classes separated by intervening, less prevalent size classes. Dune and beach sands exhibit the best sorting; river and shallow marine sands are less well sorted. River-floodplain, deltaic, and turbidity-current sand deposits show much poorer sorting. Skewness (the symmetry or asymmetry of a grain-size distribution) also varies as a function of depositional setting. Beach sands are commonly negatively skewed (they have a tail of more poorly sorted coarse grains), whereas dune and river sands tend to be positively skewed (a tail of more poorly sorted fine grains).

Careful analysis of grain roundness and grain shape also can aid in distinguishing the high-abrasion environments of beach and especially dune sands from those of fluvial or marine sands. Rounding takes place much more rapidly in sands subjected to wind action than in water-laid sands. In general, coarser sand grains are better rounded than finer grains because the coarser ones hit bottom more frequently and also hit with greater impact during transport. Sand grains may also have polished, frosted, pitted, or otherwise characteristic surfaces. These depend on the grain size, the agent of transport, and the amount of chemical attack. For example, polish can occur on medium-grained beach sands and fine-grained desert sands and can also be produced chemically by weathering processes.

Classification of Sandstones

There are many different systems of classifying sandstones, but the most commonly used schemes incorporate both texture (the presence and amount of either interstitial matrix—i.e., clasts with diameters finer than 0.03 millimetre—or chemical cement) and mineralogy (the relative amount of quartz and the relative abundance of rock fragments to feldspar grains). The system presented here is that of the American petrologist Robert H. Dott (1964), which is based on the concepts of P.D. Krynine and F.J. Pettijohn. Another popular classification is that of R.L. Folk (1974). Although these classifications were not intended to have tectonic significance, the relative proportions of quartz, feldspar, and fragments are good indicators of the tectonic regime. It is possible to discriminate between stable cratons (rich in quartz and feldspar), orogens (rich in quartz and fragments), and magmatic arcs (rich in feldspar and fragments).

Sandstones are first subdivided into two major textural groups, arenites and wackes. Arenites consist of a sand-size framework component surrounded by pore spaces that are either empty (in the case of arenite sands) or filled with crystalline chemical cement (in the case of arenites). Wackes consist of a sand-size framework component floating in a finer-grained pasty matrix of grains finer than 0.03 millimetre whose overall abundance exceeds 15 percent by volume. A third triangular panel in the background shows the natural transition from sandstones to mudrocks as the percentage of sand-size framework clasts decreases.

Further subdivision of both arenites and wackes into three specific sandstone families is based on the relative proportions of three major framework grain types: quartz (Q), feldspar (F), and rock fragments (R for rock fragment, or L for lithic fragment). For example, quartz arenites are rocks whose sand grains consist of at least 95 percent quartz. If the sand grains consist of more than 25 percent feldspar (and feldspar grains are in excess of rock fragments), the rock is termed arkosic arenite or "arkose," although such sandstones are also somewhat loosely referred to as feldspathic sandstones. In subarkosic arenite (or subarkose), feldspar sand grains likewise exceed rock fragments but range in abundance from 5 to 15 percent. Lithic arenites have rock fragments that exceed feldspar grains; the abundance of rock fragments is greater than 25 percent. Sublithic arenites likewise contain more rock fragments than feldspar, but the amount of rock fragments is lower, ranging from 5 to 25 percent. Lithic arenites can be further subdivided according to the nature of the rock fragments, as shown in the smaller triangle of figure. This classification scheme also recognizes three major types of wackes or graywackes that are roughly analogous with the three major arenite groups: quartz wacke, feldspathic wacke (with the subvariety arkosic wacke), and lithic wacke. The three major arenite sandstone families are separately described below, but the varieties of wacke can be conveniently considered together as a single group.

Quartz Arenites

Quartz arenites are usually white, but they may be any other colour; cementation by hematite, for example, makes them red. They are usually well sorted and well rounded (supermature) and often represent ancient dune, beach, or shallow marine deposits. Characteristically, they are ripple-marked or cross-bedded and occur as widespread thin blanket sands. On chemical analysis, some are found to contain more than 99 percent SiO_2 (quartz). Most commonly they are cemented with quartz, but calcite and iron oxide frequently serve as cements as well.

This type of sandstone is widespread in stable areas of continents surrounding the craton, such as central North America (St. Peter Sandstone of Ordovician age [about 505 to 438 million years old]), central Australia, or the Russian Platform, and are particularly common in Paleozoic strata (that formed from 570 to 245 million years ago). Quartz arenites have formed in the past when large areas of subcontinental dimensions were tectonically stable (not subject to uplift or deformation) and of low relief, so that extensive weathering could take place, accompanied by prolonged abrasion and sorting. This process eliminated all the unstable or readily decomposed minerals such as feldspar or rock fragments and concentrated pure quartz together with trace amounts of zircon, tourmaline, and various other resistant heavy minerals.

Quartz arenites have also accumulated to thicknesses of hundreds and even thousands of metres on the continental shelf areas produced as passive continental margins develop during the early stages of continental rifting and the opening of an ocean basin. These thick, continental margin deposits form only if source areas are sufficiently stable to permit beach abrasion and intense chemical weathering capable of destroying rock fragments and feldspars. Subsequent ocean basin closure and continental collision deforms the continental shelf and rise assemblages, incorporating clean quartz arenite units into the resulting folded and faulted mountain system, typically as major ridges. Examples include the Cambrian Chilhowee Group and Silurian Tuscarora Sandstone and Clinch Sandstone formations in the Appalachian Mountains of eastern North America and the Flathead Sandstone and Tapeats Sandstone of the Rocky Mountains in the western part of the continent.

Arkosic Sandstones

Arkosic sandstones are of two types. The most common of these is a mixture of quartz, potash feldspar, and granitic rock fragments. Chemically, these rocks are 60–70 percent silica (or silicon dioxide) and 10–15 percent aluminum oxide (Al_2O_3), with significant amounts of potassium (K), sodium (Na), and other elements. This type of arkosic sandstone, or arkose, can form wherever block faulting of granitic rocks occurs, given rates of uplift, erosion, and deposition that are so great that chemical weathering is outweighed and feldspar can survive in a relatively unaltered state. These rocks are usually reddish, generally immature, very poorly sorted, and frequently interbedded with arkose conglomerate; alluvial fans or fluvial aprons are the main depositional environments. The Triassic Newark Group of Connecticut is a classic example of this type of arkosic sandstone.

Arkoses also form under desert (or rarely Arctic) conditions in which the rate of chemical decomposition of the parent granite or gneiss is very slow. These arkoses are generally well sorted and rounded (supermature) and show other desert features, such as eolian cross-beds, associated gypsum, and other evaporitic minerals. The Precambrian Torridonian Arkose of Great Britain is thought to be of desert origin. Basal sands deposited on a granitic-gneissic craton also are usually arkosic. Subarkose sandstones (e.g., Millstone Grit from the Carboniferous of England) have a feldspar content that is diminished by more extensive weathering or abrasion or by dilution from nonigneous source rocks.

Lithic Arenites

Lithic arenites occur in several subvarieties, but they are normally gray or of salt-and-pepper appearance because of the inclusion of dark-coloured rock fragments. Most commonly, fragments of metamorphic rocks such as slate, phyllite, or schist predominate, producing phyllarenite. If volcanic rock fragments such as andesite and basalt are most abundant, the rock is termed a volcanic arenite. If chert and carbonate rock fragments are predominant, the name chert or calclithite is applied.

Lithic arenites are usually rich in mica and texturally immature; the silicon dioxide content is 60–70 percent; aluminum oxide is 15 percent; and potassium, sodium, iron (Fe), calcium (Ca), and magnesium (Mg) are present in lesser amounts. Lithic arenites are very common in the geologic record, are widespread geographically, and are of all ages. They generally were formed as the result of rapid uplift, intense erosion, and high rates of deposition. Many of the classic postorogenic clastic wedge systems found in the major mountain systems of the world contain abundant lithic arenites. In the Appalachians, these include the Ordovician Juniata Formation of the Taconic clastic wedge, the Devonian Catskill Formation of the Acadian clastic wedge, and the Pocono and Mauch Chunk formations of the Alleghenian clastic wedge. Most lithic arenites are deposited as fluvial apron, deltaic, coastal plain and shallow marine sandstones, interbedded with great thicknesses of shale and frequently with beds of coal or limestone. If they are deposited in an oxidizing environment such as a well-drained river system, they are reddish (e.g., the Catskill Formation of the northeastern United States and the Devonian Old Red Sandstone of England).

Wackes

Wacke, or graywacke, is the name applied to generally dark-coloured, very strongly bonded sandstones that consist of a heterogeneous mixture of rock fragments, feldspar, and quartz of sand size, together with appreciable amounts of mud matrix. Almost all wackes originated in the sea, and many were deposited in deep water by turbidity currents.

Interbedded shales and wackes in sedimentary rock of the Silurian Period, Lower Silurian Aberystwyth Grit Formation, Wales.

Wackes typically are poorly sorted, and the grain sizes present range over three orders of magnitude—e.g., from 2 to 2,000 micrometres ($8×10-5$ to $8×10-2$ inch). Commonly, the coarsest part of a wacke bed is its base, where pebbles may be abundant. Shale fragments, which represent lumps of mud eroded from bottom sediments by the depositing current, may be concentrated elsewhere in the bed.

Many wackes contain much mud, typically 15–40 percent, and this increases as the mean grain size of the rock decreases. The particles forming the rock are typically angular. This, and the presence of the interstitial mud matrix, has led to these rocks being called "microbreccias." The fabric and texture indicate that the sediments were carried only a short distance and were subject to very little reworking by currents after deposition.

A bed of very coarse Upper Cambrian graywacke,
showing graded bedding and load casting, Denison Range, Tasmania.

The most widespread internal structure of wackes is graded bedding, although some sequences display it poorly. Sets of cross strata more than three centimetres thick are rare, but thinner sets are very common. Parallel lamination is widespread, and convolute bedding is usually present. These internal structures are arranged within wacke beds in a regular sequence. They appear to result from the action of a single current flow and are related to changes in the hydraulics of the depositing current. In some beds, the upper part of the sequence of structures is missing, presumably because of erosion or nondeposition. In others, the lower part is missing. This has been attributed to change in the hydraulic properties of the depositing current as it moves away from its source and its velocity decreases to the point at which the first sediment deposited is laminated, rather than massive and graded as is the case closer to the source.

Groove molds on the underside of a graywacke bed from the Middle Silurian, Denbigh Grits, Wales.

The most typical external structures of wacke beds are sole markings, which occur on their undersurfaces. Flute and groove molds are the most characteristic, but many other structures have been recorded.

The upper surfaces of wacke beds are less well characterized by sedimentary structures. The most typical are current lineation and various worm tracks, particularly of the highly sinuous form

Nereites. Apart from these trace fossils, wackes are usually sparsely fossiliferous. Where fossils occur they are generally free-floating organisms (graptolites, foraminiferans) that have settled to the bottom, or bottom-living (benthic), shallow-water organisms displaced into deeper water as part of the sediment mass.

Wackes are chemically homogeneous and are generally rich in aluminum oxide (Al_2O_3), ferrous oxide (FeO)[+] ferric oxide (Fe_2O_3), magnesium oxide (MgO), and soda (Na_2O). The abundance of soda relative to potash (K_2O) (reflecting a typically high sodium plagioclase feldspar content) and dominance of ferrous oxide over ferric oxide (reflecting large amounts of chlorite in the matrix) chemically distinguishes wackes from the three arenite families. The bulk composition of most wackes mimics that of their source owing to a lack of chemical differentiation by weathering and sorting. The matrix component, which is by definition any clasts 30 micrometres or finer, allows wackes to be differentiated from the other major sandstones. To be characterized as a wacke, its matrix component must equal or exceed 15 percent; in some cases more than 50 percent matrix has been reported. The origin of the matrix component, however, is controversial. Even though laboratory studies demonstrate that gravity-driven, bottom-hugging turbidity currents deposit sand-size grains together with mud-size clasts, modern deep-sea fan and abyssal plain sands (turbidites) have a matrix component that seldom exceeds 10 percent. A large portion of the matrix in ancient wackes must therefore be secondary, derived either from the disaggregation of feldspar and fine-grained lithic fragments like shale, phyllite, and volcanic rocks or from the postdepositional infiltration of clay- and silt-size clasts from overlying beds.

Wackes are widespread in the geologic record and occur throughout geologic history. They typically are not found in association with sedimentary rocks that accumulate upon stable continental blocks and are instead confined either to intensely deformed mountain systems or to their modern analogues: ocean trenches, the continental slope and rise, and abyssal plain areas. Many, perhaps most, wackes are redeposited marine sands derived from source areas in which weathering, erosion, and deposition are too rapid to permit chemical differentiation and the breakdown of unstable components. Wackes of Archean age (those formed from 3.8 to 2.5 billion years ago) constitute the dominant sandstone type in the classic greenstone belts of the Precambrian shields (large areas of basement rocks in a craton that formed 3.8 billion to 570 million years ago around which younger sedimentary rocks have been deposited). They probably accumulated in rapidly subsiding trenches and ocean basins that surrounded primitive continental blocks. Proterozoic wackes (those formed from about 2.5 billion to 570 million years ago) are dominantly trench and ocean basin deposits, as are wackes of Phanerozoic age (those formed from 570 million years ago to the present day). They represent the accumulation of sand-size prisms of material that today are deposited both within ocean trenches (e.g., the modern trenches off Indonesia) and as submarine fan aprons (e.g., the Astoria Fan off the Pacific coast of Washington and Oregon in the United States) developed at the base of the continental slope at the mouths of submarine canyons. More distal carpets of wacke sand can extend for thousands of square kilometres across oceanic abyssal plains. Classic examples of the continental margin and ocean basin deposits include the late Precambrian Ocoee Supergroup and Ordovician Martinsburg Formation of the Appalachians, the Jurassic and Cretaceous Franciscan Formation of the Pacific Coast Ranges of California, much of the Alpine flysch of Switzerland and France, and many of the famous turbidite sands found in the Italian Apennines.

The feature common to all modern depositional sites is that they adjoin landmasses in areas of high submarine relief. The landmass may be a continent bordered by either a passive, aseismic margin (for example, the eastern margin of North America) or a seismically active margin like that found along the western coast of both North and South America. The landmass can also be an active volcanic arc such as the Aleutian Islands chain or the Japan islands arc. The critical factor is the close proximity of topographically high and emergent clastic source areas and steeply sloped submarine depositional slopes, basins, or trenches.

Mudrocks

In terms of volume, mudrocks are by far the most important variety of sedimentary rock, probably constituting nearly 80 percent of the Earth's sedimentary rock column. Despite this abundance, the literature on mudrocks does not match in extent or detail that dealing with sandstones, carbonate rocks, and the various rarer sedimentary rock varieties like evaporite and phosphorite. This paradox reflects the difficulties inherent both in analyzing such rocks, owing to their poor exposure and fine grain size, and in interpreting any data obtained from their analysis because of the effects of diagenesis. Mudrocks include all siliciclastic sedimentary rocks composed of silt- and clay-size particles: siltstone ($\frac{1}{16}$ millimetre to $\frac{1}{256}$ millimetre diameters), claystone (less than $\frac{1}{256}$ millimetre), and mudstone (a mix of silt and clay). Shale refers specifically to mudrocks that regularly exhibit lamination or fissility or both. Mudrocks are also loosely referred to as both lutites and pelites and as argillaceous sedimentary rocks.

Though mudrocks are composed mainly of detritus weathered from preexisting rocks, many contain large amounts of chemically precipitated cement (either calcium carbonate or silica), as well as abundant organic material. Mudrocks produced from the alteration of volcanic lava flows and ash beds to clay and zeolite minerals are called bentonites.

General Properties of Shales

The properties of shales are largely determined by the fine grain size of the constituent minerals. The accumulation of fine clastic detritus generally requires a sedimentary environment of low mechanical energy (one in which wave and current actions are minimal), although some fine material may be trapped by plants or deposited as weakly coherent pellets in more agitated environments. The properties of the clay mineral constituents of lutites are particularly important, even when they do not make up the bulk of a rock.

The mineralogy of shales is highly variable. In addition to clay minerals (60 percent), the average shale contains quartz and other forms of silica, notably amorphous silica and cristobalite (30 percent), feldspars (5 percent), and the carbonate minerals calcite and dolomite (5 percent). Iron oxides and organic matter (about 0.5 and 1 percent, respectively) are also important. Older estimates greatly underestimated clay minerals because of incorrect assignment of potassium to feldspar minerals. The most abundant clay mineral is illite; montmorillonite and mixed-layer illite-montmorillonite are next in abundance, followed by kaolinite, chlorite, chlorite-montmorillonite, and vermiculite. The quartz-to-feldspar ratio generally mirrors that of associated sands. In pelagic (deep-sea) sediments, however, feldspar may be derived from local volcanic sources, whereas quartz may be introduced from the continents by wind, upsetting simple patterns. A large number of accessory minerals occur in shales. Some of these are detrital, but diagenetic or in situ varieties e.g., pyrite, siderite, and various phosphates and volcanically derived varieties.

Origin of Shales

The formation of fine-grained sediments generally requires weak transporting currents and a quiet depositional basin. Water is the common transporting medium, but ice-rafted glacial flour (silt produced by glacial grinding) is a major component in high-latitude oceanic muds, and wind-blown dust is prominent, particularly in the open ocean at low and intermediate latitudes. Shale environments thus include the deep ocean; the continental slope and rise; the deeper and more protected parts of shelves, shallow seas, and bays; coastal lagoons; interdistributory regions of deltas, swamps, and lakes (including arid basin playas); and river floodplains. The deep-sea muds are very fine, but an orderly sequence from coarse sediments in high-energy nearshore environments to fine sediments at greater depths is rarely found. Sediments at the outer edges of present-day continental shelves are commonly sands, relict deposits of shallower Pleistocene (from about 2.6 million to 11,700 years ago) glacial conditions, whereas muds are currently being deposited in many parts of the inner shelf. The nearshore deposition of clay minerals is enhanced by the tendency of riverborne dispersed platelets to flocculate in saline waters (salinity greater than about four parts per thousand) and to be deposited just beyond the agitated estuarine environment as aggregates hydraulically equivalent to coarser particles. Differential flocculation leads to clay-mineral segregation, with illite and kaolinite near shore and montmorillonite farther out to sea. Advance of silty and sandy delta-slope deposits over clays also leads to complex grain-size patterns.

Shales may be deposited in environments of periodic agitation. Sediments deposited on submarine slopes are frequently mechanically unstable and may be redistributed by slumping and turbidity currents to form thick accumulations (possible present-day eugeosynclinal equivalents) on the lower continental slope and rise. Part of the shale in many wacke-shale alternations may be of turbidite origin. Fine sediment can be deposited in marshes and on tidal flats. Trapping by marsh plants and binding of muds in fecal pellets are important. Because of electrochemical interactions among fine particles, muds plastered on a tidal flat by an advancing tide are difficult to reerode on the ebb. This may lead, as in the present-day Waddenzee, in The Netherlands, to a size increase from nearshore tidal flat muds to lag sands seaward. Fine floodplain sediments may dry out to coherent shale pellets, and these, on reerosion, can be redistributed as sands and gravels.

Shales of Economic Value

Black shales are often of economic importance as sources of petroleum products and metals, and this importance will probably increase in the future. The lacustrine Eocene Green River Shales of Colorado, Wyoming, and Utah are potentially rich petroleum sources and are undergoing exploratory extraction. Bituminous layers of the Early Permian Irati Shales of Brazil are similarly important. These shales contain the remains of the marine reptile Mesosaurus, also found in South Africa, and played a prominent part in the development of the concepts of continental drift. The widespread thin Chattanooga Shale (Devonian-Mississippian) of the eastern United States has been exploited for its high (up to 250 parts per million) uranium content. The Kupferschiefer of the Permian (286 to 245 million years old) is a bituminous shale rich in metallic sulfides of primary sedimentary or early diagenetic origin; it covers a large area of central Europe as a band generally less than one metre thick, and in eastern Germany and in Poland there is sufficient enrichment in copper, lead, and zinc for its exploitation as an ore.

Limestones and Dolomites

Limestones and dolomites are collectively referred to as carbonates because they consist predominantly of the carbonate minerals calcite ($CaCO_3$) and dolomite ($CaMg[CO_3]_2$). Almost all dolomites are believed to be produced by recrystallization of preexisting limestones, although the exact details of this dolomitization process continue to be debated. Consequently, the following discussion initially deals with limestones and dolomites as a single rock type and subsequently considers the complex process by which some limestones become dolomite.

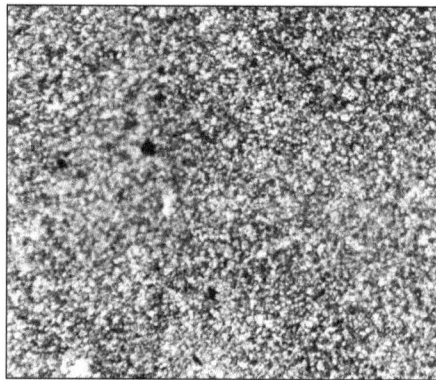

Photomicrograph showing micritic limestone from the Triassic Period (magnified 18×).

Carbonates are by far the only volumetrically important nonsiliciclastic sedimentary rock type. Most are marine, and thick sequences of carbonate rocks occur in all the continental blocks, a surviving record of the transgressions and regressions of shallow marine (epeiric) seas that repeatedly blanketed the stable continental cratonic areas from time to time mainly during the late Precambrian, Paleozoic, and Mesozoic eras. Modern marine carbonate sediments, whose formation is favoured by warm, shallow water, are presently being deposited in a broad band straddling the Equator. The texture, sedimentary structures, composition, and organic content of carbonates provide numerous insights into the environment of deposition and regional paleogeography. Many important oil reservoirs of the world, especially those of the Middle East, occur in carbonate rocks.

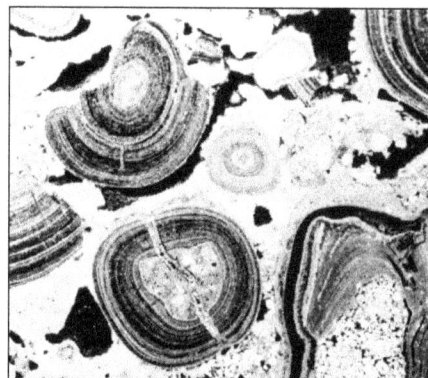

Photomicrograph showing pisolitic dolomite from the Upper Triassic Period (magnified 5×).

Mineralogy

Though ancient limestones and dolomites are composed of calcite and dolomite, respectively, other calcite group minerals such as magnesite ($MgCO_3$), rhodochrosite ($MnCO_3$), and siderite ($FeCO_3$) occur in limited amounts in restricted environments. Modern carbonate sediments are composed almost entirely of metastable aragonite ($CaCO_3$) and magnesium-rich calcite, both of which readily recrystallize during diagenesis to form calcite. Carbonate rocks commonly grade naturally into siliciclastic sedimentary rocks as the proportion of terrigenous grains of varying size and mineralogy increases. Such mixtures are the consequence of the infringement of a dominantly siliciclastic depositional setting (e.g., a quartz arenitic beach area) into, for example, a lagoon or tidal flat in which carbonate mud accumulates.

Textural Components

Carbonate minerals present in ancient limestones and dolomites occur in one of three textural forms: (1) discrete silt to sand to coarser carbonate grains, or allochems, such as oöids or skeletal fragments, (2) mud-size interstitial calcium carbonate matrix called microcrystalline calcite or micrite, and (3) interlocking, 0.02- to 0.1-millimetre-diameter crystals of clear interstitial calcium carbonate cement or spar.

In a rather simplistic sense, these three carbonate rock textural components are comparable, respectively, to the three possible constituents in a sandstone: (1) the coarser rock and mineral grains, (2) interstitial matrix, and (3) interstitial chemical cement.

Several types of allochems exist: oöids, skeletal grains, carbonate clasts, and pellets. Oöids (also known as oölites or oöliths) are sand-size spheres of calcium carbonate mud concentrically laminated about some sort of nucleus grain, perhaps a fossil fragment or a silt-size detrital quartz grain. Oöids develop today on shallow shelf areas where strong bottom currents can wash the various kinds of material that form oöid nuclei back and forth in well-agitated, warm water that is supersaturated with calcium carbonate. The concentric layers of aragonite (in modern oöids) is produced by blue-green algae that affix themselves to the grain nucleus. Skeletal fragments, also known as bioclasts, can be whole fossils or broken fragments of organisms, depending on current and wave strength as well as depositional depth. The content and texture of the bioclast component in any carbonate will vary noticeably as a function of both age (due to evolution) and depositional setting (because of subsequent abrasion and transport as well as ecology). Carbonate clasts include fragments weathered from carbonate source rocks outside the depositional basin (lithoclasts) as well as fragments of carbonate sediment eroded from within the basin almost immediately after it was deposited (intraclasts). Silt- to sand-size particles of microcrystalline calcite or aragonite that lack the internal structure of oöids or bioclasts generally are called pellets or peloids. Most are fecal pellets generated by mud-ingesting organisms. Pellets can be cemented together into irregularly shaped composite grains dubbed lumpstones or grapestones.

Microcrystalline carbonate mud (micrite) and sparry carbonate cement (sparite) are collectively referred to as orthochemical carbonate because, in contrast to allochems, neither exhibits a history of transport and deposition as clastic material. Micrite can occur either as matrix that fills or partly fills the interstitial pores between allochems or as the main component of a carbonate rock. It originates mainly as the result of organic activity: algae generate tiny needles of aragonite within their tissues, and after their death such needles fall to the depositional surface as unconsolidated mud, which soon recrystallizes to calcite. Some micrite is produced by inorganic precipitation of aragonite; grain-to-grain collision and the resulting abrasion of allochems also can generate modest amounts of micrite. Most of the coarser and clearer crystals of sparry calcite that fill interstitial pores as cement represent either recrystallized micrite or essentially a direct inorganic precipitate.

A number of carbonate classification schemes have been developed, but most modern ones subdivide and name carbonate rock types on the basis of the kinds of allochems present and the nature of the interstitial pore filling, whether it is micrite or spar. The most widely used scheme of this type is the descriptive classification devised by the American petrologist Robert L. Folk.

Origin of Limestones

Limestones originate mainly through the lithification of loose carbonate sediments. Modern carbonate sediments are generated in a variety of environments: continental, marine, and transitional, but most are marine. The present-day Bahama banks is the best known modern carbonate setting. It is a broad submarine shelf covered by shallow, warm seawater. The Bahama shelf, or carbonate platform, mimics the setting that repeatedly prevailed across the stable cratonic areas of the major continental blocks during late Precambrian, Paleozoic, and Mesozoic time and serves as a model for explaining the various limestone types that make up such ancient carbonate successions.

The edge of the shelf is marked by a topographically sharp escarpment flanked by coarse, angular limestone breccia. Submarine channels etched into the escarpment serve as waterways down which shallow-water carbonate sediment can be transported by turbidity currents capable of redistributing them as apronlike deposits on the oceanic abyssal plain. In many areas, the fringe of the Bahama banks is marked by wave-resistant reef rocks (sometimes classified as boundstone). Abrasion of these reefs by wave activity generates abundant skeletal debris. Variations in depth and current strength control the relative amounts of micrite and sparite, the prevalence of specific organisms and their productivity, and the likelihood of generating oöids, pellets, and carbonate rock fragments. Micrite and micritic allochemical sediments accumulate in deep-water, low-energy, protected areas like lagoons and tidal flats and on the leeward side of major islands. In high-energy, shallow-water locales such as beaches, coastal dunes, and tidal channels, currents winnow out any micrite, and these become the sites of sparry allochemical sediment deposition. Pinpointing the exact depositional setting for an ancient carbonate deposit requires detailed analysis of its texture, composition, sedimentary structures, geometry, fossil content, and stratigraphic relationships with modern carbonate depositional sites.

In addition to the ancient analogues of the modern carbonate deposits described above are freshwater limestones (marls) and limestone muds (or calcilutites) of deep-water abyssal plains. Freshwater limestones of limited extent represent a spectrum of small-scale settings developed within and along the margins of lacustrine basins. Deep-water abyssal plain limestones are quite restricted in volume and age in the geologic record for a number of reasons. First of all, abyssal plain sequences are less likely to be incorporated into the orogenic belts that develop as continental margins are compressed during ocean basin closure. Second, pelagic calcareous oozes are the obvious modern analogues of ancient abyssal plain calcilutites. These oozes are produced by aragonite-secreting plankton that float near the surface (such as foraminiferans and coccoliths), which upon their death leave their shells, or tests, to settle slowly to the ocean bottom and accumulate. The development of such deep-sea deposits is therefore obviously dependent on the existence of calcium-secreting planktonic organisms, and these did not evolve until Mesozoic time. Finally, calcareous ooze accumulation is severely restricted both by latitude (being largely confined to a band extending 30° to 40° north and south of the Equator) and abyssal plain depth (approximately 2,000 metres). Below a depth of about 4,500 metres, which is the carbonate compensation depth (CCD), the pressure and temperature of seawater produces a rate of dissolution in excess of the rate of pelagic test accumulation.

Dolomites and Dolomitization

Dolomite is produced by dolomitization, a diagenetic process in which the calcium carbonate minerals aragonite and calcite are recrystallized and converted into the mineral dolomite. Dolomitization can obscure or even obliterate all or part of the original limestone textures and structures; in the case where such original features survive, carbonate nomenclature and interpretation can still be applied to the rock with emphasis on the effects of alteration.

The exact processes by which limestones are dolomitized are not thoroughly understood, but dolomites occur widely in the geologic record. The relative proportion of dolomite to limestone progressively increases with age in carbonate rocks. This secular trend probably either reflects the earlier existence of geochemical settings that were more favourable to dolomitization or is the logical result of the fact that the likelihood for a limestone to undergo dolomitization increases proportionally with its age.

Geochemists have been unable to precipitate normal dolomite under the conditions of temperature and pressure that exist in nature; temperatures within the 200 °C range are required to support precipitation. A few modern, so-called primary marine dolomite localities have been studied, but close investigation of these areas suggests that even these penecontemporaneous dolomites are produced by altering calcite or aragonite almost immediately after their initial precipitation. Dolomites generated by later alteration of older limestones are known as diagenetic dolomites.

The study of the few reported penecontemporaneous dolomite sites allows some conclusions to be formed regarding the dolomitization process. These modern dolomites develop mainly under conditions of high salinity (hypersalinity), which commonly exist in arid regions across supratidal mud flats as well as on the flat, saline plains and playa lake beds known as sabkhas. In highly saline environments, the ratio of dissolved magnesium ions to dissolved calcium ions progressively increases above the norm for seawater (5:1) as a result of the selective formation of calcium-rich evaporite minerals like gypsum and anhydrite. These magnesium-rich brines then tend to be flushed downward owing to their high density; the entire process is named evaporative reflux. Penecontemporaneous dolomites would result from the positioning of sabkhas and arid supratidal flats in a site that is in immediate contact with carbonate sediment; diagenetic dolomites would logically result when such dolomite-producing settings overlie older limestone deposits. The presence of fissures or highly permeable zones serving as channelways for downward percolation of dolomitizing fluids would also promote the alteration. Other studies have emphasized a possible role in dolomitization for dense brackish (salty) fluids formed when seawater and meteoric waters (those precipitated from the atmosphere as rain or snow) are produced along coastal zones.

Siliceous Rocks

Those siliceous rocks composed of an exceptionally high amount of crystalline siliceous material, mainly the mineral quartz (especially microcrystalline quartz and fibrous chalcedony) and amorphous opal, are most commonly known as chert. A wide variety of rock names are applied to cherty rocks reflecting their colour (flint is dark chert; jasper is usually red; prase is green) and geographic origin (novaculite of Arkansas, U.S.; silexite of France). The term chert is applied here to all fine-grained siliceous sediments and sedimentary rocks of chemical, biochemical, and organic origin.

Types of Cherts

Two major varieties of chert deposits exist—namely, bedded chert and nodular chert. Bedded cherts occur in individual bands or layers ranging in thickness from one to several centimetres or even tens of metres. They are intimately associated with volcanic rocks, commonly submarine volcanic flows as well as deep-water mudrocks. Classic examples include the Miocene Monterey Formation of the Coast Ranges of California, the Permian Rex Chert of Utah and Wyoming, the Arkansas Novaculite of the Ouachita Mountains, and the Mesozoic chert deposits of the Franciscan Formation of California. Nodular cherts occur as small to large (millimetres to centimetres) knotlike and fistlike clusters of quartz, chalcedony, and opal concentrated along or parallel with bedding planes in shallow-water marine carbonate rocks as well as pelagic limestones. Individual nodules may be ovoid or semispherical in shape; masses of chert typically form an anastamosing network.

Origin of Cherts

Many bedded cherts are composed almost entirely of the remains of silica-secreting organisms like diatoms and radiolarians. Such deposits are produced by compacting and recrystallizing the organically produced siliceous ooze deposits that accumulate on the present-day abyssal ocean floor. The modern oozes gather in latitudes where high organic productivity of floating planktonic radiolarians and diatoms takes place in the warm surface waters. As individual organisms die, their shells settle slowly to the abyssal floor and accumulate as unconsolidated siliceous ooze. Siliceous oozes are particularly prominent across areas of the ocean floor located far from continental blocks, where the rate of terrigenous sediment supply is low, and in deeper parts of the abyssal plain lying below the carbonate compensation depth, where the accumulation of calcareous oozes cannot occur. Some bedded cherts might not be of organic origin. They instead may be produced by precipitating silica gels derived from the same magma chambers from which the submarine basalts (pillow lava) that are intimately associated with bedded cherts are precipitated.

The origin of nodular cherts has long been debated, but most are produced by the secondary replacement of the carbonate minerals and fossils within shallow marine shelf deposits. Evidence of secondary origin includes relict structures of allochems such as skeletal fragments and oöids preserved entirely within chert nodules. Silica can be mobilized from elsewhere within a rock and transported in solution under proper conditions of temperature and geochemistry. Likely sources of silica found scattered within shallow-water shelf carbonates include siliceous sponge spicules, radiolarians or diatom shells, and windblown sand grains. The details of the process and the possible role of microscopic organisms like bacteria in dissolving, mobilizing, and reconcentrating the silica remains uncertain.

Finally, geysers and hot springs like those of the Yellowstone National Park area of northwestern Wyoming, U.S., are also sites of chert deposition. Encrustations of silica, known as sinter or geyserite, are volumetrically unimpressive but nevertheless are curiosities. The geyser and hot springs activity at Yellowstone is probably typical, with a subterranean body of magma as the source of silica-rich hydrothermal solutions rising periodically near or to the surface.

Phosphorites

Many sedimentary rocks contain phosphate in the form of scattered bones composed of the mineral apatite (calcium phosphate), but rocks composed predominantly of phosphate are rare. Nevertheless, three principal types exist: (1) regionally extensive, crystalline nodular, and bedded phosphorites, (2) localized concentrations of phosphate-rich clastic deposits (bone beds), and (3) guano deposits.

Evaporites

Evaporites are layered crystalline sedimentary rocks that form from brines generated in areas where the amount of water lost by evaporation exceeds the total amount of water from rainfall and influx via rivers and streams. The mineralogy of evaporite rocks is complex, with almost 100 varieties possible, but less than a dozen species are volumetrically important. Minerals in evaporite rocks include carbonates (especially calcite, dolomite, magnesite, and aragonite), sulfates (anhydrite and gypsum), and chlorides (particularly halite, sylvite, and carnallite), as well as various borates, silicates, nitrates, and sulfocarbonates. Evaporite deposits occur in both marine and nonmarine sedimentary successions.

Though restricted in area, modern evaporites contribute to genetic models for explaining ancient evaporite deposits. Modern evaporites are limited to arid regions (those of high temperature and low rates of precipitation), for example, on the floors of semidry ephemeral playa lakes in the Great Basin of Nevada and California, across the coastal salt flats (sabkhas) of the Middle East, and in salt pans, estuaries, and lagoons around the Gulf of Suez. Ancient evaporates occur widely in the Phanerozoic geologic record, particularly in those of Cambrian (from 570 to 505 million years ago), Permian (from 286 to 245 million years ago), and Triassic (from 245 to 208 million years ago) age, but are rare in sedimentary sequences of Precambrian age. They tend to be closely associated with shallow marine shelf carbonates and fine (typically rich in iron oxide) mudrocks. Because evaporite sedimentation requires a specific climate and basin setting, their presence in time and space clearly constrains inferences of paleoclimatology and paleogeography. Evaporite beds tend to concentrate and facilitate major thrust fault horizons, so their presence is of particular interest to structural geologists. Evaporites also have economic significance as a source of salts and fertilizer.

All evaporite deposits result from the precipitation of brines generated by evaporation. Laboratory experiments can accurately trace the evolution of brines as various evaporite minerals crystallize. Normal seawater has a salinity of 3.5 percent (or 35,000 parts per million), with the most important dissolved constituents being sodium and chlorine. When seawater volume is reduced to one-fifth of the original, evaporite precipitation commences in an orderly fashion, with the more insoluble components (gypsum and anhydrite) forming first. When the solution reaches one-tenth the volume of the original, more soluble minerals like sylvite and halite form. Natural evaporite sequences show vertical changes in mineralogy that crudely correspond to the orderly appearance of mineralogy as a function of solubility but are less systematic.

Nonmarine Environment

Evaporite deposition in the nonmarine environment occurs in closed lakes—i.e., those without outlet—in arid and semiarid regions. Such lakes form in closed interior basins or shallow depressions on land where drainage is internal and runoff does not reach the sea. If water depths are shallow or, more typically, somewhat ephemeral, the term playa or playa lake is commonly used.

Water inflow into closed lakes consists principally of precipitation and surface runoff, both of which are small in amount and variable in occurrence in arid regions. Groundwater flow and discharge from springs may provide additional water input, but evaporation rates are always in excess of precipitation and surface runoff. Sporadic or seasonal storms may give rise to a sudden surge of water inflow. Because closed lakes lack outlets, they can respond to such circumstances only by deepening and expanding. Subsequent evaporation will reduce the volume of water present to prestorm or normal amount; fluctuation of closed lake levels therefore characterizes the environment.

Such changing lake levels and water volumes lead to fluctuating salinity values. Variations in salinity effect equilibrium relations between the resulting brines and lead to much solution and subsequent reprecipitation of evaporites in the nonmarine environment. As a result of these complexities as well as the distinctive nature of dissolved constituents in closed lake settings, nonmarine evaporite deposits contain many minerals that are uncommon in marine evaporites—e.g., borax, epsomite, trona, and mirabilite.

Shallow Marine Environment

Evaporite deposition in the shallow marine environment (sometimes termed the salina) occurs in desert coastal areas, particularly along the margins of such semi-restricted water bodies as the Red Sea, Persian Gulf, and Gulf of California. Restriction is, in general, one of the critical requirements for evaporite deposition, because free and unlimited mixing with the open sea would allow the bodies of water to easily overcome the high evaporation rates of arid areas and dilute these waters to near-normal salinity. This semi-restriction cannot, in fact, prevent a large amount of dilution by mixing; coastal physiography is the principal factor involved in brine production. Shallow-water evaporites, almost exclusively gypsum, anhydrite, and halite, typically interfinger with tidal flat limestone and dolomite and fine-grained mudrock.

Deep-basin Environment

Most of the thick, laterally extensive evaporite deposits appear to have been produced in deep, isolated basins that developed during episodes of global aridity. The most crucial requirement for evaporite production is aridity; water must be evaporated more rapidly than it can be replenished by precipitation and inflow. In addition, the evaporite basin must somehow be isolated or at least partially isolated from the open ocean so that brines produced through evaporation are prevented from returning there. Restricting brines to such an isolated basin over a period of time enables them to be concentrated to the point where evaporite mineral precipitation occurs. Periodic breaching of the barrier, due either to crustal downwarping or to global sea-level changes, refills the basin from time to time, thereby replenishing the volume of seawater to be evaporated and making possible the inordinately thick, regionally extensive evaporite sequences visible in the geologic record.

Debate continues over the exact mechanisms for generating thick evaporite deposits. Three possible models for restricting "barred" evaporite basins are shown in Figure. They differ in detail, and none has garnered a consensus of support. The deep-water, deep-basin model accounts for replenishment of the basin across the barrier or sill, with slow, continual buildup of thick evaporites made possible by the seaward escape of brine that allows a constant brine concentration to be maintained. The shallow-water, shallow-basin model produces thick evaporites by continual subsidence of the basin floor. The shallow-water, deep-basin model shows the brine level in the basin beneath the level of the sea as a result of evaporation; brines are replenished by groundwater recharge from the open ocean.

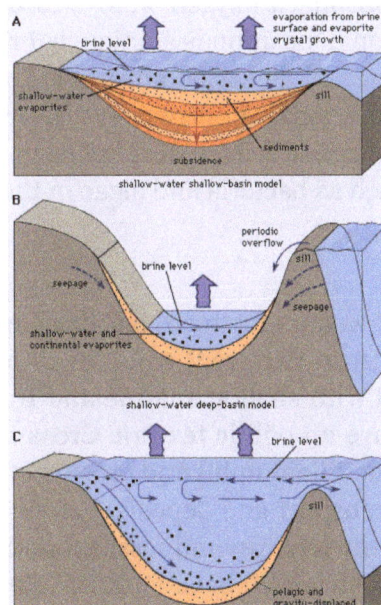

Three models for deposition of marine evaporites in basins of restricted water circulation.

Iron-rich Sedimentary Rocks

Almost all sedimentary rocks are iron-bearing in the sense that mudrocks, sandstones, and carbonates typically have an iron content of several percent. Nevertheless, sedimentary rocks in which the proportion of iron exceeds 15 percent are separately categorized as iron-rich. Two major types of iron-rich sedimentary rocks are recognized: (1) iron formation, or banded iron formation (BIF)—regionally extensive, locally thick sequences composed of alternating thin (millimetre to centimetre thick) layers of mainly crystalline-textured iron-rich minerals and chert—and (2) ironstone—noncherty, essentially clastic-textured, iron-rich minerals of local extent.

Banded iron formations are predominantly Precambrian in age; most are 1.8 to 2.2 billion years old; none are younger than Cambrian age. The most important iron-bearing minerals in iron formations are hematite, magnetite, and greenalite. These deposits constitute the world's major source of iron ore. Classic examples are found in the Mesabi Range of Minnesota, U.S., and the Kiruna ores of Sweden.

Ironstones are principally of Phanerozoic age, mainly Early Paleozoic (roughly 440 to 570 million years old) and Jurassic (about 144 to 208 million years old), but can be as old as Middle Precambrian age (about 1.6 to 3 billion years old). They appear to be restricted to basins no larger than 150 kilometres in any direction. Major iron minerals are goethite, hematite, and chamosite.

Origin of Banded Iron Formation

The origin of banded iron formation is not clearly understood. Banded iron formation units are typically 50 to 600 metres thick. Their complex mineralogy includes various iron oxides, iron carbonates, iron silicates, and iron sulfides. The essentially crystalline texture of these minerals together with the definitive crystalline texture of the laminated chert bands with which the iron mineral layers alternate is perplexing. At present, iron is not easily dissolved, nor can it be readily transported in solution and subsequently precipitated as crystalline-textured, iron-rich minerals, because of the presence of free atmospheric oxygen. Many sedimentary petrologists consequently conclude that banded iron formation deposition is a uniquely Precambrian occurrence made possible by, and supporting the existence of, an earlier anaerobic Earth atmosphere (one lacking free oxygen) quite unlike that in existence today. Controversy also continues over the ultimate source of iron (weathering as opposed to magmatic iron escaping from the Earth's interior) and over the possible role of microorganisms such as bacteria and algae in the precipitation of the iron.

Origin of Ironstones

The origin of ironstones also is not well understood, but most appear to be derived from the erosion and redeposition of lateritic (iron-rich, red) soils. Ironstones occur as thin (a few tens of metres at most) units interbedded with shallow marine and transitional carbonates, mudrocks, and sandstones. They generally have an oölitic texture. Cross-bedding, ripple marks, and small scour and fill channels are abundant. Slight uplift and erosion of reddish soils developed in coastal regions drained by rivers that transport and deposit such material in deltas and embayments along the coast is compatible with the features and fossils found in deposits of this sort. Classical ironstone deposits include the Ordovician Wabana Formation of Newfoundland and the Silurian Clinton Group of the central and southern Appalachians.

Organic-rich Sedimentary Deposits

Coal, oil shale, and petroleum are not sedimentary rocks per se; they represent accumulations of undecayed organic tissue that can either make up the bulk of the material (e.g., coal), or be disseminated in the pores within mudrocks, sandstones, and carbonates (e.g., oil shale and petroleum). Much of the undecomposed organic matter in sediment and sedimentary rocks is humus, plant matter that accumulates in soil. Other important organic constituents include peat, humic organic matter that collects in bogs and swamps where oxidation and bacterial decay is incomplete, and sapropel, fine-grained organic material—mainly the soft organic tissue of phytoplankton and zooplankton, along with bits and pieces of higher plants—that amass subaqueously in lakes and oceans.

Organic-rich sedimentary rock deposits are collectively referred to as fossil fuels because they consist of the undecayed organic tissue of plants and animals preserved in depositional settings characterized by a lack of free oxygen. Fossil fuels constitute the major sources of energy in the industrial world, and their unequal distribution in time (exclusively Phanerozoic) and space (more than half of the proved petroleum reserves are in the Persian Gulf region of the Middle East) has a significant effect on the world's political and economic stability.

Coal

Coals are the most abundant organic-rich sedimentary rock. They consist of undecayed organic matter that either accumulated in place or was transported from elsewhere to the depositional site. The most important organic component in coal is humus. The grade or rank of coal is determined by the percentage of carbon present. The term peat is used for the uncompacted plant matter that accumulates in bogs and brackish swamps. With increasing compaction and carbon content, peat can be transformed into the various kinds of coal: initially brown coal or lignite, then soft or bituminous coal, and finally, with metamorphism, hard or anthracite coal. In the geologic record, coal occurs in beds, called seams, which are blanketlike coal deposits a few centimetres to metres or hundreds of metres thick.

Many coal seams occur within cyclothems, rhythmic successions of sandstone, mudrock, and limestone in which nonmarine units are regularly and systematically overlain by an underclay, the coal seam itself, and then various marine lithologies. The nonmarine units are thought to constitute the floor of ancient forests and swamps developed in low-lying coastal regions; the underclay is a preserved relict of the soil in which the coal-producing vegetation was rooted; and the marine units overlying the coal record the rapid transgression of the sea inland that killed the vegetation by drowning it and preventing its decomposition by rapid burial. The exact mechanism responsible for generating the rapid episodes of marine transgression and regression necessary to generate coal-bearing cyclothems is not definitively known. A combination of episodic upwarping and downwarping of the continental blocks or global (eustatic) changes in sea level or both, coupled with normal changes in the rate of sediment supply that occurs along coasts traversed by major laterally meandering river systems, may have been the cause.

In any case, coal is a rare, though widely distributed, lithology. Extensive coal deposits overall occur mainly in rocks of Devonian age (those from 408 to 360 million years old) and younger because their existence is clearly contingent on the evolution of land plants. Nevertheless, small, scattered coal deposits as old as early Proterozoic have been described. Coal-bearing cyclothem deposits are especially abundant in the middle and late Paleozoic sequences of the Appalachians and central United States and in the Carboniferous of Great Britain, probably because during this time interval global climates were warm and humid and large portions of the continental blocks were low-lying platforms located only slightly above sea level.

Oil Shale

Mudrock containing high amounts of organic matter in the form of kerogen is known as oil shale. Kerogen is a complex waxy mixture of hydrocarbon compounds composed of algal remains or of amorphous organic matter with varying amounts of identifiable organic remnants. The most famous oil shale deposit in the world, located in the United States, is the Green River Formation of Utah, Wyoming, and Colorado of Eocene age (i.e., formed 57.8 to 36.6 million years ago). This vast deposit contains fossils and sedimentary structures, suggesting rapid deposition and burial of unoxidized organic matter in shallow lakes or marine embayments. The quantity of oil that can be extracted from the kerogen in the Green River Formation is significant. The cost of extracting the oil by heating the shale, however, far exceeds the cost of extracting equivalent quantities of crude oil, natural gas, or coal. Also, the traditional aboveground refractory process requires extensive strip mining and immense volumes of water, and even more modern methods of heating the shale underground consume large amounts of water and generate significantly more carbon dioxide than the extraction of other fossil fuels does. Despite these considerable economic and technical problems, oil shales potentially represent a significant future energy resource.

Oil and Natural Gas

Natural gas refers collectively to the various gaseous hydrocarbons generated below the Earth's surface and trapped in the pores of sedimentary rocks. Major natural gas varieties include methane, ethane, propane, and butane. These natural gases are commonly, though not invariably, intimately associated with the various liquid hydrocarbons—mainly liquid paraffins, napthenes, and aromatics—that collectively constitute oil.

Hydrocarbons can also exist in a semisolid or solid state such as asphalt, asphaltites, mineral waxes, and pyrobitumens. Bitumens can occur as seepages, impregnations filling the pore space of sediments (e.g., tar sands of the Canadian Rocky Mountains), and in veins or dikes. Asphaltites occur primarily in dikes and veins that cross sedimentary rocks such as gilsonite deposits in the Green River Formation of Utah. These natural bitumens probably form from the loss of volatiles, oxidation, and biological degradation resulting from oil seepage to the surface. Solid hydrocarbons are of interest to geologists as their presence is a good indicator of petroleum below the surface in that region. Also, solid hydrocarbons have commercial value.

The exact process by which oil and natural gas are produced is not precisely known, despite the extensive efforts made to determine the mode of petroleum genesis. Crude oil is thought to form from undecomposed organic matter, principally single-celled floating phytoplankton and zooplankton that settle to the bottom of marine basins and are rapidly buried within sequences of mudrock and limestone. Natural gas and oil are generated from such source rocks only after heating and compaction. Typical petroleum formation (maturation) temperatures do not exceed 100 °C, meaning that the depth of burial of source rocks cannot be greater than a few kilometres. After their formation, oil and natural gas migrate from source rocks to reservoir rocks composed of sedimentary rocks largely as a consequence of the lower density of the hydrocarbon fluids and gases. Good reservoir rocks, by implication, must possess high porosity and permeability. A high proportion of open pore spaces enhances the capacity of a reservoir to store the migrating petroleum; the interconnectedness of the pores facilitates the withdrawal of the petroleum once the reservoir rock is penetrated by drill holes.

Secular Trends in the Sedimentary Rock Record

Reexamination of the sedimentary rock record preserved within the continental blocks suggests systematic changes through time in the relative proportions of the major sedimentary rock types deposited. These changes can be linked to the evolution of the atmosphere and hydrosphere and to the changing global tectonic setting. Carbonates and quartz sands, for example, require long-term source area stability as well as the existence of broad, shallow-water epeiric seas that mantle continental blocks. Marine transgressions and regressions across broad stable continental cratons occurred only in Proterozoic and Phanerozoic time; Archean continental blocks were smaller and tectonically unstable, and most likely less granitic than those of today. Consequently, the early Precambrian sedimentary rock record consists largely of volcanogenic sediments, wackes, and arkoses physically disintegrated from small, high-relief island arcs (the Archean greenstone belts of the various Precambrian shields) and microcontinental fragments. The fact that iron formations are restricted to rocks of Archean and Proterozoic time supports the conclusion that atmospheric oxygen levels in earlier stages of Earth history were lower, promoting the dissolution, transport, and precipitation of iron by chemical or biochemical means. The total lack of evaporites from the Archean record and their subsequent steady buildup probably reflects a number of factors. Such deposits can be easily destroyed by metamorphism, and presumably, given enough time, they will have been completely erased. Also, evaporite formation requires seawater with elevated salinity, a condition that is established with time. Finally, significant volumes of bedded evaporites can occur only once broad, stable continents have evolved, because the requisite restricted evaporite basins develop exclusively adjacent to cratons.

Metamorphic Rocks

Metamorphic rocks are basically rocks that have experience change due to high pressure and temperature below zone of diagenesis. Protolith refers to the original rock, prior to metamorphism. In low grade metamorphic rocks, original textures are often preserved allowing one to determine the likely protolith. As the grade of metamorphism increases, original textures are replaced with metamorphic textures and other clues, such as bulk chemical composition of the rock, are used to determine the protolith. Below is an examination of the role of two agents of metamorphism

The Role of Temperature

Changes in temperature conditions during metamorphism cause several important processes to occur. With increasing temperature, and thus higher energy, chemical bonds are able to break and reform driving the chemical reactions that changes the rock's chemistry during metamorphism. Increasing in temperature can also result in the growth of crystals. In a rock, a small number of large crystals have a higher thermodynamic stability than do a large number of small crystals. As a result, increasing temperature during metamorphism, even in the absence of any chemical change, will generally result in the amalgamation of small crystals to produce a coarser grained rock. It is a fact that individual minerals are only stable over specific temperature ranges. Thus, as temperature changes, minerals within a rock become unstable and transform through chemical reactions to new minerals. This property is very important to our interpretation of metamorphic rocks. By observing the mineral assemblage (set of minerals) within a metamorphic rock, it is often possible to make an estimate of the temperature at the time of formation. That is, minerals can be used as thermometers of the process of metamorphism.

The Role of Pressure

Pressure, the second of the two physical parameters controlling metamorphism and occurs in two forms. The most widely experienced type of pressure is lithostatic. This "rock-constant" pressure is derived from the weight of overlying rocks. Lithostatic pressure is experienced uniformly by a metamorphic rock. That is, the rock is squeezed to the same degree in all directions. Thus, there is no preferred orientation to lithostatic pressure and there is no mechanical drive to rearrange crystals within a metamorphic rock experiencing lithostatic conditions. The second pressure is the directed pressure, this is pressure of motion and action. Plate tectonics provide the underlying mechanical control for all forms of directed pressure. Thus, metamorphism is closely linked to the plate tectonic cycle and many metamorphic rocks are the products of tectonic interactions. As was the case with changes in temperature, changes in pressure, either lithostatic or directed, have important impacts upon the stability of minerals. Every mineral is stable over a range of pressures, if pressure conditions during metamorphism exceed a mineral's stability range the mineral will transform to a new phase. Many of these solid-state reactions involve polymorphic transformation – changes between minerals with the same chemistry and different crystallographic structures. Just as with temperature, mineral assemblages within a metamorphic rock can be used as a barometer to measure pressure at the time of formation.

Classification

Classification of metamorphic rocks depends on textures and its degree of metamorphism. Three kinds of criteria are normally employed in the classification of metamorphic rock. These are:

- Mineralogical - The most abundant minerals are used as a prefix to a textural term. Thus, a schist containing biotite, garnet, quartz, and feldspar, would be called a biotite-garnet schist. A gneiss containing hornblende, pyroxene, quartz, and feldspar would be called a hornblende-pyroxene gneiss. A schist containing porphyroblasts of K-feldspar would be called a K-spar porphyroblastic schist.

- Chemical - If the general chemical composition can be determined from the mineral assemblage, then a chemical name can be employed. For example a schist with a lot of quartz and feldspar and some garnet and muscovite would be called a garnet-muscovite quartzo-feldspathic schist. A schist consisting mostly of talc would be called a talc-magnesian schist.

- Texture- Most metamorphic textures involve foliation. Foliation is generally caused by a preferred orientation of sheet silicates. If a rock has a slaty cleavage as its foliation, it is termed a slate, if it has a phyllitic foliation, it is termed a phyllite, if it has a shistose foliation, and it is termed a schist. A rock that shows a banded texture without a distinct foliation is termed a gneiss. All of these could beporphyroblastic (i.e. could contain porhyroblasts).A rock that shows no foliation is called a hornfels if the grain size is small, and a granulite, if the grain size is large and individual minerals can be easily distinguished with a hand lens.

Metamorphic Grade

The intensity of a metamorphic event through the use of the concept of metamorphic grade (Figure below). With increasing depth in the Earth, ambient temperature and pressure conditions rise steadily. Thus, within the continental crust, temperatures vary from approximately 200 °C at 5 km to 800 °C at 35 km. While these temperatures are extreme relative to our everyday experiences, they are significantly below the melting point of most rocks. Likewise, lithostatic pressure increases with increasing depth. At 5 km the pressure is approximately 2 kilo bars, or about 2000 times atmospheric pressure. Deeper within the crust, at about 35 km, the pressure increases to some 10 kb. This trend of increasing temperature and pressure within the Earth is defined by a region of commonly encountered metamorphic conditions. Low temperature and pressure setting as low-grade metamorphism usually gneisses, while high temperature and intense pressure is known as high-grade metamorphism in schist environment.

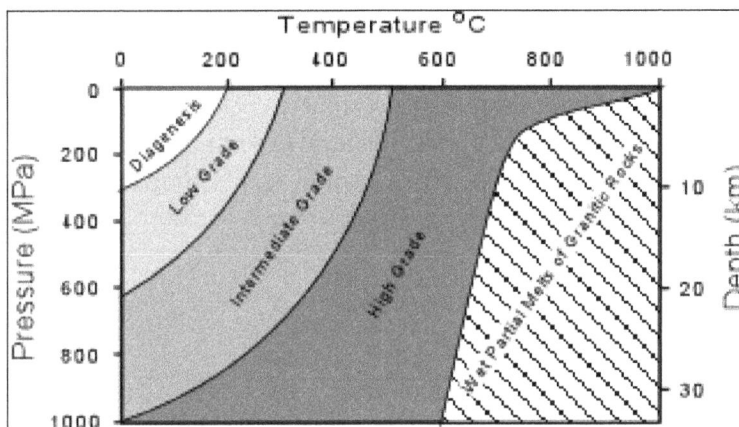

Grade of metamorphism gneiss to schist.

Types of Metamorphism

Contact Metamorphism

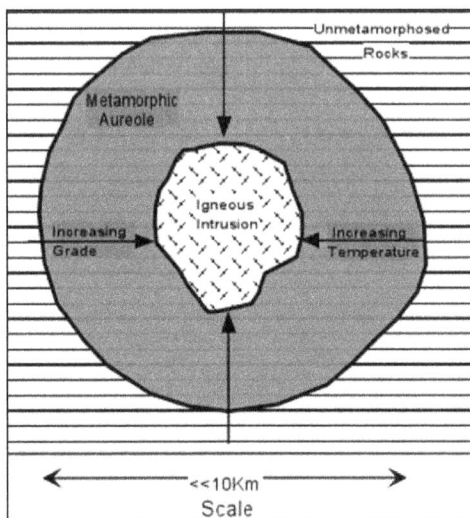

Metamorphic aureole.

Contact metamorphism occurs adjacent to igneous intrusions and results from high temperatures associated with the igneous intrusion. Since only a small area surrounding the intrusion is heated by the magma, metamorphism is restricted to the zone surrounding the intrusion, called a metamorphic or contact aureole. Outside of the contact aureole, the rocks are not affected by the intrusive event. The grade of metamorphism increases in all directions toward the intrusion. Because the temperature contrast between the surrounding rock and the intruded magma is larger at shallow levels in the crust where pressure is low, contact metamorphism is often referred to as high temperature, low pressure metamorphism. The rock produced is often a fine-grained rock that shows no foliation, called a hornfels.

Regional Metamorphism

Regional metamorphism occurs over large areas and generally does not show any relationship to igneous bodies. Most regional metamorphism is accompanied by deformation under non-hydrostatic or differential stress conditions. Thus, regional metamorphism usually results in forming metamorphic rocks that are strongly foliated, such as slates, schists, and gneisses. The differential stress usually results from tectonic forces that produce compressional stresses in the rocks, such as when two continental masses collide. Thus, regionally metamorphosed rocks occur in the cores of fold/thrust mountain belts or in eroded mountain ranges. Compressive stresses result in folding of rock and thickening of the crust, which tends to push rocks to deeper levels where they are subjected to higher temperatures and pressures.

Cataclastic Metamorphism

Cataclastic metamorphism occurs as a result of mechanical deformation, like when two bodies of rock slide past one another along a fault zone. Heat is generated by the friction of sliding along such a shear zone, and the rocks tend to be mechanically deformed, being crushed and pulverized, due to the shearing. Cataclastic metamorphism is not very common and is restricted to a narrow zone along which the shearing occurred.

Hydrothermal Metamorphism

Rocks that are altered at high temperatures and moderate pressures by hydrothermal fluids are hydrothermally metamorphosed. This is common in basaltic rocks that generally lack hydrous minerals. The hydrothermal metamorphism results in alteration to such Mg-Fe rich hydrous minerals as talc, chlorite, serpentine, actinolite, tremolite, zeolites, and clay minerals. Rich ore deposits are often formed as a result of hydrothermal metamorphism.

Burial Metamorphism

When sedimentary rocks are buried to depths of several hundred meters, temperatures greater than 300oC may develop in the absence of differential stress. New minerals grow, but the rock does not appear to be metamorphosed. The main minerals produced are often the Zeolites. Burial metamorphism overlaps, to some extent, with diagenesis, and grades into regional metamorphism as temperature and pressure increase.

Shock Metamorphism (Impact Metamorphism)

When an extra-terrestrial body, such as a meteorite or comet impacts with the Earth or if there is a very large volcanic explosion, ultrahigh pressures can be generated in the impacted rock. These ultrahigh pressures can produce minerals that are only stable at very high pressure, such as the SiO_2 polymorphs coesite and stishovite. In addition they can produce textures known as shock lamellae in mineral grains, and such textures as shatter cones in the impacted rock.

Metamorphic Facies

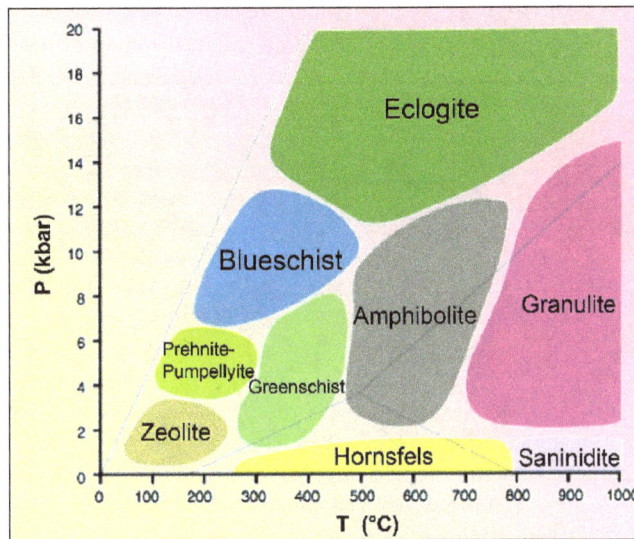

Metamorphic faces.

The changes in mineral assemblages are due to changes in the temperature and pressure conditions of metamorphism. Thus, the mineral assemblages that are observed must be an indication of the temperature and pressure environment that the rock was subjected to. This pressure and temperature environment is referred to as Metamorphic Facies. The sequence of metamorphic facies observed in any metamorphic terrain, depends on the geothermal gradient that was present during metamorphism. Each facies has specific index minerals.

Zeolite zeolites: especially laumontite, wairakite, analcime

Prehnite-Pumpellyite prehnite + pumpellyite (+ chlorite + albite)

Greenschist chlorite + albite + epidote (or zoisite) + quartz ± actinolite

Amphibolite hornblende + plagioclase (oligoclase-andesine) ± garnet

Granulite orthopyroxene (+ clinopyrixene + plagioclase ± garnet ± hornblende)

Blueschist glaucophane + lawsonite or epidote (+albite ± chlorite)

Eclogite pyrope garnet + omphacitic pyroxene (± kyanite)

Contact facies

Igneous Activity, Metamorphism and Plate Tectonics

Plate tectonics is the mechanisms behind motion of crustal plates. It is still a controversial hypothesis though it explains very well volcanicity and igneous activity. Through this mechanism heat is transferred from deeper levels of the earth through plate margins. At these environments igneous activities and metamorphism is widespread. i.e. convergent plate margins and divergent plate tectonic margins. Convergent tectonic margins are areas around ring of fire covering Indonesia, Philippines and South America in the Andes region.

Divergent plate margins are the mid-Atlantic ridge and the African Rift Valley. At these margins major geothermal resources occur e.g. in Philippines, Indonesia, Iceland and African countries within Rift valley.

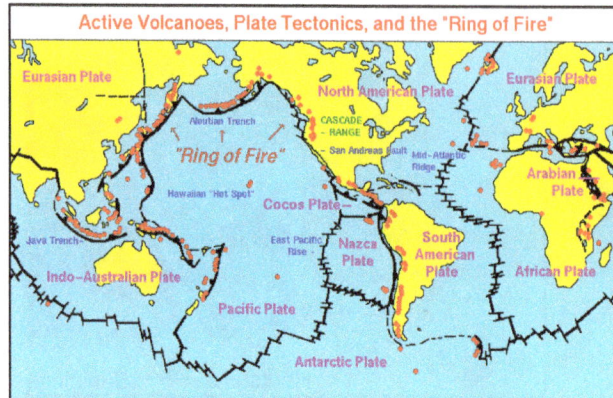

Plate tectonics and igneous provinces.

Divergent Plate Boundary

Earth plates are made of rigid lithosphere and the plastic asthenosphere. In the divergent boundaries usually basic igneous rock types are produced. Asthenosphere is usually close to the surface in the order of 5 to 10 km. It is usually plastic since temperatures are lower than those required for melting. When this zone is upwelled the pressure is reduced and melting occurs due to decompression from deeper levels hence melting. Beneath the divergent boundary the asthenosphere is welled upward and decompression occurs. At this environments Igneous of volcanic activities are pretty common and with the temperatures and pressure metamorphism also is with spread.

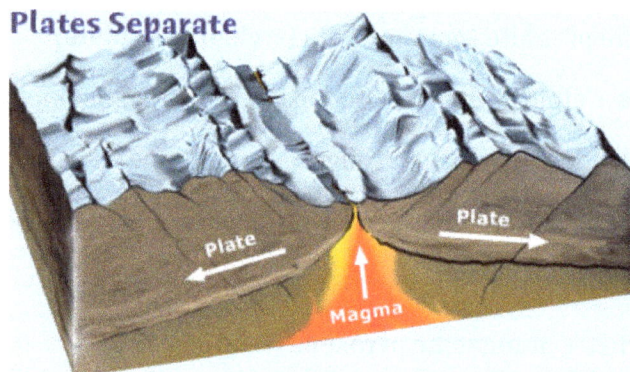

Divergent plate boundary.

Intraplate Zones

Igneous activity away from plate margin is unusual however, such hotspots exist due to mantle plumes e.g. in Hawaii, and yellow stone national park.

Convergent Plates

Intermediate and silic magmas are clearly related to convergence plate and subduction.

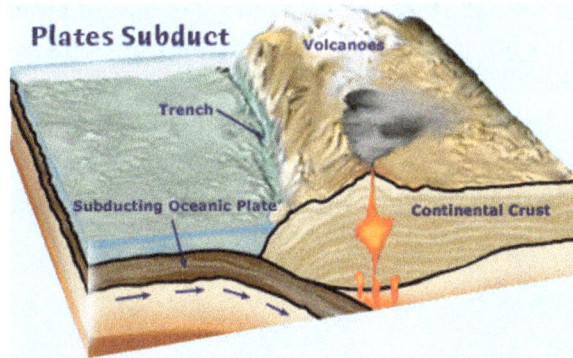

Convergent Plate Boundary

Geothermal Resources

Different types of geothermal resources occur, namely. Those associated with volcanicity or igneous environments, sedimentary/geopressurised environments and to lesser extent metamorphic or dry rock system. The systems associated with igneous or volcanic environments are by far more and are high enthalpy resources making them very attractive for development. Example of such resources are found in Kenya within igneous provinces of the Rift Valley, Iceland, Hawaii etc. Only in a few places like Hawaii are intraplate activities associated with geothermal are found.

Rock Cycle

The rock components of the crust are slowly but constantly being changed from one form to another and the processes involved are summarized in the rock cycle (Figure below). The rock cycle is driven by two forces: (1) Earth's internal heat engine, which moves material around in the core and the mantle and leads to slow but significant changes within the crust, and (2) the hydrological cycle, which is the movement of water, ice, and air at the surface, and is powered by the sun.

The rock cycle is still active on Earth because our core is hot enough to keep the mantle moving, our atmosphere is relatively thick, and we have liquid water. On some other planets or their satellites, such as the Moon, the rock cycle is virtually dead because the core is no longer hot enough to drive mantle convection and there is no atmosphere or liquid water.

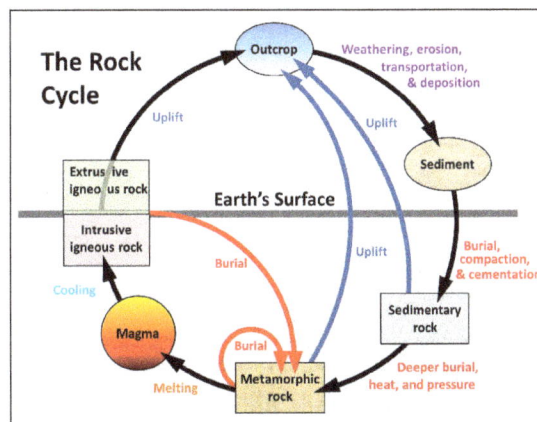

A schematic view of the rock cycle.

In describing the rock cycle, we can start anywhere we like, although it's convenient to start with magma. magma is rock that is hot to the point of being entirely molten. This happens at between about 800° and 1300°C, depending on the composition and the pressure, onto the surface and cool quickly (within seconds to years) — forming extrusive igneous rock.

Magma forming pahoehoe basalt at Kilauea Volcano.

Magma can either cool slowly within the crust (over centuries to millions of years) — forming intrusive igneous rock, or erupt onto the surface and cool quickly (within seconds to years) — forming extrusive igneous rock. Intrusive igneous rock typically crystallizes at depths of hundreds of metres to tens of kilometres below the surface. To change its position in the rock cycle, intrusive igneous rock has to be uplifted and exposed by the erosion of the overlying rocks.

Cretaceous-aged marine sandstone overlying mudstone.

Through the various plate-tectonics-related processes of mountain building, all types of rocks are uplifted and exposed at the surface. Once exposed, they are weathered, both physically (by mechanical breaking of the rock) and chemically (by weathering of the minerals), and the weathering products — mostly small rock and mineral fragments — are eroded, transported, and then deposited as sediments. Transportation and deposition occur through the action of glaciers, streams, waves, wind, and other agents, and sediments are deposited in rivers, lakes, deserts, and the ocean.

Metamorphosed and folded Triassic-aged limestone.

Unless they are re-eroded and moved along, sediments will eventually be buried by more sediments. At depths of hundreds of metres or more, they become compressed and cemented into sedimentary rock. Again through various means, largely resulting from plate-tectonic forces, different kinds of rocks are either uplifted, to be re-eroded, or buried deeper within the crust where they are heated up, squeezed, and changed into metamorphic rock.

References

- What-are-minerals, rocks-and-minerals, subsoil: eniscuola.net, Retrieved 4 July, 2019

- What-is-a-mineral, minerals: geology.com, Retrieved 14 April, 2019

- Minerals: rsc.org, Retrieved 18 February, 2019

- Properties-of-minerals: rocksandminerals4u.com, Retrieved 1 July, 2019

- Major-minerals-characteristics, geology: qsstudy.com, Retrieved 9 May, 2019

- Rock-geology, science: britannica.com, Retrieved 2 March, 2019

- Igneous-rock, science: britannica.com, Retrieved 22 August, 2019

- Sedimentary-rock, science: britannica.com, Retrieved 30 June, 2019

- The-rock-cycle, geology: opentextbc.ca, Retrieved 17 January, 2019

Chapter 4

Geomorphic Processes

Geomorphic processes are the endogenic and exogenic forces which cause chemical actions and physical stresses on earth materials. Some of the geomorphic processes are mass movements, erosion and deposition, and soil formation. The topics elaborated in this chapter will help in gaining a better perspective about these types of geomorphic processes.

Endogenic and exogenic forces that cause physical stresses and chemical actions upon earth materials and bring about changes in the surface configuration of the earth are known as Geomorphic Processes. The Earth's surface is continuously exposed to endogenic as well as exogenic forces.

Gradation – Phenomenon of wearing down of relief variations of surface of earth through erosion. Exogenic forces lead to either degradation or aggradation.

- Degradation : results in wearing down of relief or elevation

- Aggradation: results in filling up of basins or depressions

Geomorphic agents – Any exogenic element of nature which is capable of acquiring and transporting earth material is an agent. Examples, water, ice, wind.

Role of Gravity and Gradient

- Gravity is essentially a directional force which activates all down-slope movements of matter, and causes stresses on materials.

- Transportation and deposition of materials is possible due to gravity and gradients, otherwise there would be no mobility and erosion.

- Indirect gravitational stresses lead to activation of wave and tide induced currents and winds.

- Therefore, gravitational stresses are as important as the other geomorphic processes.

- Definition: Gravity is the force that keeps us in contact with the surface and it is the force which triggers the movement of all surface material on earth.

- Also, all the movements either within the earth, or over surface of the earth, occur due to gradients — from higher levels to lower levels, or from high pressure to lower pressure areas, etc.

Endogenic Processes

These are processes emanating from the interior of Earth and induce diastrophism and volcanism in the lithosphere.

- Endogenic forces – are mainly land building forces. They arise from radioactivity, rotational friction, tidal friction and primordial heat from the origin of the earth.
- Distribution – variations in geothermal gradients, heat flow from inside of earth, crustal thickness and strength, cause the action of endogenic forces be non-uniform. Hence, the tectonically controlled original crustal surface is uneven.

Diastrophism

All processes that involve moving, elevating or building up components of the earth's crust are categorized as diastrophism.

These processes are studied under following heads:

(i) Orogenic processes:

- Mountain building through folding.
- It affects long and narrow belts of the earth's crust.
- Crust is deformed in form of folds.

(ii) Epeirogenic processes:

- Continent building process.
- Involves simple deformation of crust.
- Under this, uplift or warping of large parts of the earth's crust occurs.

(iii) Earthquakes involve relatively local and minor movements.

(iv) Plate tectonics involve horizontal movements of crustal plates.

All the above four processes exhibit following characteristics which induce metamorphism of rocks:

- Faulting and fracturing of crust occurs.
- Pressure temperature and volume (PVT) changes occur.

Volcanism

The movement of molten rock (magma) over, or towards the earth's surface while also forming various intrusive and extrusive volcanic forms.

Exogenic Process

These processes are mainly land wearing processes. They derive their energy from atmospheric sources including the Sun and gradients of tectonic factors.

- Gravitational stresses: Forces acting along earth materials are sheer stresses (force applied per unit area). It breaks rocks and boulders. Shear stress results in angular displacement or slippage.
- Molecular stresses: they occur due to temperature changes, crystallization and melting. Climatic processes that control various processes are mainly- temperature and precipitation.

Weathering

Weathering is defined as mechanical disintegration and chemical decomposition of rocks through the actions of various elements of weather and climate. It is in-situ (on site) process. Climate is main factor, also topography and vegetation.

There are three types of weathering processes: (i) Chemical; (ii) Physical/ Mechanical; (iii) Biological weathering processes.

Chemical Weathering

- Water and air (oxygen and carbon dioxide) and heat.

Solution-

$$CO_2 + H_2O \rightarrow H_2CO_3$$

$$CACO_3 + H_2CO_3 \rightarrow Ca(HCO_3)_2$$

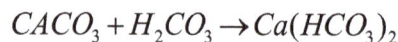

- Soluble rock-forming minerals like nitrates, sulphates and potassium are dissolved in water from solid and disintegrate.

- These leave rainy climates and accumulate in dry regions and areas.

- Calcium carbonate and calcium bicarbonate are present in limestone which soluble in carbonic acid (carbon dioxide and water).

- Sodium chloride is also susceptible to solubility.

Carbonation

- Reaction of carbonate and bicarbonate with minerals and breaks down feldspar and carbonate minerals.

- Calcium carbonates and magnesium carbonates are dissolved in carbonic acid and these are removed in solution without residue resulting in cave formation.

Hydration

- Chemical addition of water.
- Minerals take up water and expand.
- Calcium sulfate takes in water and turns to gypsum which is more unstable than calcium sulfate.
- It is reversible process.

Oxidation and Reduction

- Oxidation is combination of a mineral with oxygen to form oxides or hydroxide.
- When ready access to atmospheric oxygen and water.
- Iron, manganese, sulphur, etc.
- Breakdown of minerals occurs due to the disturbance caused by addition of oxygen.
- Red colour of iron, on oxidation turns brown or yellow, and on reduction turns to greenish grey.
- When oxidized minerals are exposed to an environment where oxygen is absent, reduction takes place.

Physical Weathering

It is caused by thermal expansion and pressure release. when repeated, cause continued fatigue of the rock.

Applied Forces

- Gravitational forces: overburden pressure, load and shearing stress.
- Expansion forces: due to temperature changes, crystal growth or animal activity.
- Water pressures are controlled by wetting and drying cycles.

Unloading and Expansion

- Exfoliation sheets resulting from expansion due to unloading and pressure release may measure hundreds or even thousands of meters in horizontal extent.

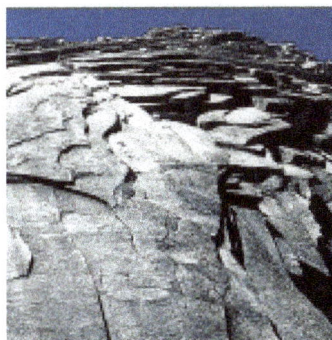

- Fractures develop roughly parallel to the ground surface.

- Large, smooth rounded domes formed due to unloading called exfoliation domes.

Temperature Changes and Exapansion

- This is most effective and evident in dry climates and high elevations where diurnal temperature changes are drastic.

- Surface layer expands more than the rock beneath and leads to formation of stress within rock resulting in heaving and fracturing parallel to square.

- In granites, tors (smooth surfaced and small rounded) form due to such exfoliation that occurs due to thermal expansion.

- Freezing, thawing and frost wedging

- Due to growth of ice within pores and cracks of rocks during repeated cycles of freezing and melting.

- High elevations in mid-latitudes.

- Glacial areas have frost wedging daily.

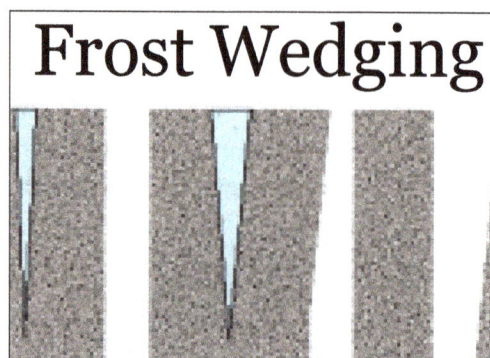

Salt Weathering

- Salts expand due to thermal action, hydration, crystallization.

- Expansion depend on temperature and their thermal properties.

- Calcium, sodium , magnesium , potassium, barium have tendency to expand.

- 30-50 degree Celsius surface temperature in deserts is favorable.

- Granular disintegration or foliation is observed in salt weathering.

- Salt crystallization is most effective in this category.

- Areas with alternate wetting and drying conditions favor salt crystal growth.

- Chalk breaks most readily, then Limestone, sandstone, shale, gnesiss, and granite, etc.

Biological Weathering

- Due to growth and movement of organisms.

- Burrowing, wedging by earthworms, termites, rodents, etc.

- Expose new surfaces to chemical attack.

- Decaying plants and animals matter produce humic, carbonic acid.

- Plant roots exert pressure mechanically.

Mass Movements

Mass movement, also called Mass Wasting, are bulk movements of soil and rock debris down slopes in response to the pull of gravity, or the rapid or gradual sinking of the Earth's ground surface in a predominantly vertical direction. Formerly, the term mass wasting referred to a variety of processes by which large masses of crustal materials are moved by gravity from one place to another. More recently, the term mass movement has been substituted to include mass wasting processes and the sinking of confined areas of the Earth's ground surface. Mass movements on slopes and sinking mass movements are often aided by water and the significance of both types is the part each plays in the alteration of landforms.

The variety of downslope mass movements reflects the diversity of factors that are responsible for their origin. Such factors include: weathering or erosional debris cover on slopes, which is usually liable to mass movement; the character and structure of rocks, such as resistant permeable beds prone to sliding because of underlying impermeable rocks; the removal of the vegetation cover, which increases the slope's susceptibility to mass movement by reducing its stability; artificial or natural increases in the slope's steepness, which will usually induce mass movement; earthquake tremors, which affect the slope equilibrium and increase the likelihood of mass movement; and flowing ground water, which exerts pressure on soil particles and impairs slope stability. These factors affecting slope conditions will often combine with climatic factors such as precipitation and frost activity to produce downslope mass movement.

The types of mass movements caused by the above factors include: the abrupt movement and free fall of loosened blocks of solid rock, known as rock-falls; several types of almost imperceptible downslope movement of surficial soil particles and rock debris, collectively called creep; the subsurface creep of rock material, known as bulging: the multiplicity of downslope movements of bedrock and other debris caused by the separation of a slope section along a plane of least resistance or slip surface, collectively called landslides; the separation of a mass along a concave head scarp, moving down a curved slip surface and accumulating at the slope's foot, known as a slump; the saturation of debris and weathered material by rainfall in the upper section of a slope or valley, increasing the weight of the debris and causing a slow downslope movement, called an earthflow; a rapidly moving earthflow possessing a higher water content, known as a mudflow; a fast-moving earthflow in a mountainous region, called a debris flow or avalanche; and the downslope movement of moisture-saturated surficial material, known as solifluction, over frozen substratum material, occurring in sub-Arctic regions during seasonal periods of surface thaw.

Sinking mass movements occur in relatively rapid fashion, known as subsidence, and in a gradual manner, called settlement. Subsidence involves a roof collapse or breakdown of a subsurface cavity such as a cave. Extensive subsidence is evident in areas where coal, salt, and metalliferous ores are mined. Marine erosion sometimes causes the roof collapse of sea caves. Regions of karst topography will exhibit widespread subsidence in the form of sinkholes caused by underground drainage. Other types of subsidence caused by underground solutions have been found in chalk, gypsum, anhydrite, halite (salt), and loess terrains. The melting of ground ice also contributes to subsidence such as the formation of glacial kettles and depressions following the seasonal surface thaw of perennially frozen land. The chemical decomposition of subsurface rocks and ores is also a cause of subsidence. Another form of subsidence is the steep-walled depression, known as a volcanic sink, formed following the withdrawal of magma from below the ground surface.

The gradual settlement of confined areas of earth material occurs through consolidation of soil and rock by the squeezing or removal of fluids from the pore spaces, and by the collapse of the grain structure. The most widespread cause of consolidation is by surface loading such as the continued deposition of sediments in sea and lake beds or by loads imposed on land by glacial ice sheets or outwash deposits. Human-made structures also cause surface loading, consolidation, and settlement. Consolidation is also caused by the lowering of the ground water table. The extraction of pressurized water or oil from deep beneath the surface will cause a collapse of the pore spaces and consolidation of rock material. Grain structure collapse usually occurs from the wetting of rock materials such as clays and sands, which causes the structure of the grains to shift and settle in a more compact and dense configuration.

Erosion and Deposition

Erosion is defined as the removal of soil, sediment, regolith, and rock fragments from the landscape. Most landscapes show obvious evidence of erosion. Erosion is responsible for the creation of hills and valleys. It removes sediments from areas that were once glaciated, shapes the shorelines of lakes and coastlines, and transports material downslope from elevated sites. In order for erosion to occur three processes must take place: detachment, entrainment and transport. Erosion also requires a medium to move material. Wind, water, and ice are the mediums primarily responsible for erosion. Finally, the process of erosion stops when the transported particles fall out of the transporting medium and settle on a surface. This process is called deposition. Illustrates an area of Death Valley, California where the effects of erosion and deposition can be easily seen.

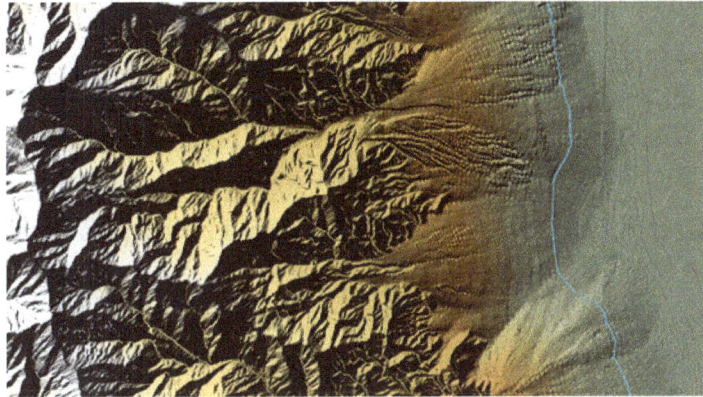

The left is the Panamint Mountain Range. To the right is Death Valley. Elevation spans from 3368 to -83meters and generally decreases from left to right. The blue line represents an elevation of 0 meters. Large alluvial fans extending from a number of mountain valleys to the floor of Death Valley can be seen in the right side of the image. The sediments that make up these depositional features came from the weathering and erosion of bedrock in the mountains located on the left side of the image.

Energy of Erosion

The energy for erosion comes from several sources. Mountain building creates a disequilibrium within the Earth's landscape because of the creation of relief. Gravity acts to vertically move materials of higher relief to lower elevations to produce an equilibrium. Gravity also acts on the mediums of erosion to cause them to flow to base level.

Solar radiation and its influence on atmospheric processes is another source of energy for erosion. Rainwater has a kinetic energy imparted to it when it falls from the atmosphere. Snow has potential energy when it is deposited in higher elevations. This potential energy can be converted into the energy of motion when the snow is converted into flowing glacial ice. Likewise, the motion of air because of differences in atmospheric pressure can erode surface material when velocities are high enough to cause particle entrainment.

The Erosion Sequence

Erosion can be seen as a sequence of three events: detachment, entrainment, and transport. These three processes are often closely related and sometimes not easy distinguished between each other. A single particle may undergo detachment, entrainment, and transport many times.

Detachment

Erosion begins with the detachment of a particle from surrounding material. Sometimes detachment requires the breaking of bonds which hold particles together. Many different types of bonds exist each with different levels of particle cohesion. Some of the strongest bonds exist between the particles found within igneous rocks. In these materials, bonds are derived from the growth of mineral crystals during cooling. In sedimentary rocks, bonds are weaker and are mainly caused by the cementing effect of compounds such as iron oxides, silica, or calcium. The particles found in soils are held together by even weaker bonds which result from the cohesion effects of water and the electro-chemical bonds found in clay and particles of organic matter.

Physical, chemical, and biological weathering act to weaken the particle bonds found in rock materials. As a result, weathered materials are normally more susceptible than unaltered rock to the forces of detachment. The agents of erosion can also exert their own forces of detachment upon the surface rocks and soil through the following mechanisms:

- Plucking: ice freezes onto the surface, particularly in cracks and crevices, and pulls fragments out from the surface of the rock.

- Cavitation: intense erosion due to the surface collapse of air bubbles found in rapid flows of water. In the implosion of the bubble, a micro-jet of water is created that travels with high speeds and great pressure producing extreme stress on a very small area of a surface. Cavitation only occurs when water has a very high velocity, and therefore its effects in nature are limited to phenomenon like high waterfalls.

- Raindrop impact: the force of a raindrop falling onto a soil or weathered rock surface is often sufficient to break weaker particle bonds. The amount of force exerted by a raindrop is a function of the terminal velocity and mass of the raindrop.

- Abrasion: the excavation of surface particles by material carried by the erosion agent. The effectiveness of this process is related to the velocity of the moving particles, their mass, and their concentration at the eroding surface. Abrasion is very active in glaciers where the particles are firmly held by ice. Abrasion can also occur from the particles held in the erosional mediums of wind and water.

Entrainment

Entrainment is the process of particle lifting by the agent of erosion. In many circumstances, it is hard to distinguish between entrainment and detachment. There are several forces that provide particles with a resistance to this process. The most important force is frictional resistance. Frictional resistance develops from the interaction between the particle to its surroundings. A number of factors increase frictional resistance, including: gravity, particle slope angle relative to the flow direction of eroding medium, particle mass, and surface roughness.

Entrainment also has to overcome the resistance that occurs because of particle cohesive bonds. These bonds are weakened by weathering or forces created by the erosion agent (abrasion, plucking, raindrop impact, and cavitation).

Entrainment Forces

The main force responsible for entrainment is fluid drag. The strength of fluid drag varies with the mass of the eroding medium (water is 9000 times more dense than air) and its velocity. Fluid drag causes the particle to move because of horizontal force and vertical lift. Within a medium of erosion, both of these forces are controlled by velocity. Horizontal force occurs from the push of the agent against the particle. If this push is sufficient to overcome friction and the resistance of cohesive bonds, the particle moves horizontally. The vertical lift is produced by turbulence or eddies within the flow that push the particle upward. Once the particle is lifted the only force resisting its transport is gravity as the forces of friction, slope angle, and cohesion are now non-existent. The particle can also be transported at velocities lower than the entrainment velocities because of the reduction in forces acting on it.

Many hydrologists and geomorphologists require a mathematical model to predict levels of entrainment, especially in stream environments. In these highly generalized models, the level of particle entrainment is relative to particle size and the velocity of the medium of erosion. These quantitative models can be represented graphically. On these graphs, the x-axis represents the log of particle diameter, and the y-axis the log of velocity. The relationship between these two variables to the entrainment of particles is described by a curve, and not by a straight line.

The critical entrainment velocity curve suggests that particles below a certain size are just as resistant to entrainment as particles with larger sizes and masses. Fine silt and clay particles tend to have higher resistance to entrainment because of the strong cohesive bonds between particles. These forces are far stronger than the forces of friction and gravity.

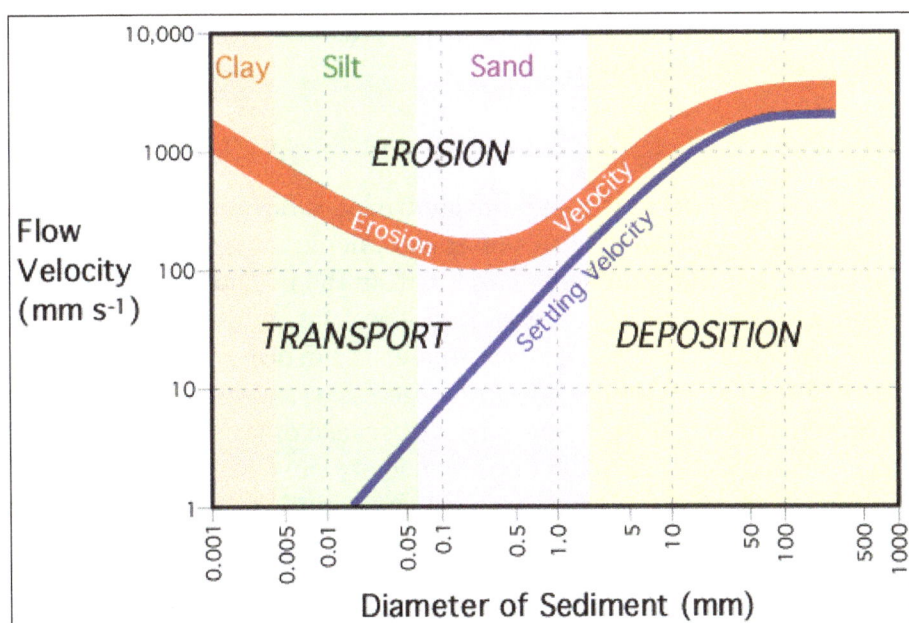

This graph describes the relationship between stream flow velocity and particle erosion, transport, and deposition. The curved line labeled "erosion velocity" describes the velocity required to entrain particles from the stream's bed and banks. The erosion velocity curve is drawn as a thick line because the erosion particles tends to be influenced by a variety of factors that changes from stream to stream. Also, note that the entrainment of silt and clay needs greater velocities then larger sand particles. This situation occurs because silt and clay have the ability to form cohesive bounds between particles. Because of the bonding, greater flow velocities are required to break the bonds and move these particles. The graph also indicates that the transport of particles requires lower flow velocities then erosion. This is especially true of silt and clay particles. Finally, the line labeled "settling velocity" shows at what velocity certain sized particles fall out of transport and are deposited.

Transport

Once a particle is entrained, it tends to move as long as the velocity of the medium is high enough to transport the particle horizontally. Within the medium, transport can occur in four different ways:

- Suspension is where the particles are carried by the medium without touching the surface of their origin. This can occur in air, water, and ice.

- Saltation is where the particle moves from the surface to the medium in quick continuous repeated cycles. The action of returning to the surface usually has enough force to cause the entrainment of new particles. This process is only active in air and water.

- Traction is the movement of particles by rolling, sliding, and shuffling along the eroded surface. This occurs in all erosional mediums.

- Solution is a transport mechanism that occurs only in aqueous environments. Solution involves the eroded material being dissolve and carried along in water as individual ions.

Particle weight, size, shape, surface configuration, and medium type are the main factors that determine which of these processes operate.

Deposition

The erosional transport of material through the landscape is rarely continuous. Instead, we find that particles may undergo repeated cycles of entrainment, transport, and deposition. Transport depends on an appropriate balance of forces within the transporting medium. A reduction in the velocity of the medium, or an increase in the resistance of the particles may upset this balance and cause deposition. Reductions in competence can occur in a variety of ways. Velocity can be reduced locally by the sheltering effect of large rocks, hills, stands of vegetation or other obstructions. Normally, competence changes occur because of large scale reductions in the velocity of flowing medium. For wind, reductions in velocity can be related to variations in spatial heating and cooling which create pressure gradients and wind. In water, lower velocities can be caused by reductions in discharge or a change in the grade of the stream. Glacial flows of ice can become slower if precipitation input is reduced or when the ice encounters melting. Deposition can also be caused by particle precipitation and flocculation. Both of these processes are active only in water. Precipitation is a process where dissolved ions become solid because of changes in the temperature or chemistry of the water. Flocculation is a chemical process where salt causes the aggregation of minute clay particles into larger masses that are too heavy to remain suspended.

Soil Formation

Soil is the thin layer of material covering the earth's surface and is formed from the weathering of rocks. It is made up mainly of mineral particles, organic materials, air, water and living organisms—all of which interact slowly yet constantly.

Most plants get their nutrients from the soil and they are the main source of food for humans, animals and birds. Therefore, most living things on land depend on soil for their existence.

Soil is a valuable resource that needs to be carefully managed as it is easily damaged, washed or blown away. If we understand soil and manage it properly, we will avoid destroying one of the essential building blocks of our environment and our food security.

The Soil Profile

As soils develop over time, layers (or horizons) form a soil profile. Most soil profiles cover the earth as 2 main layers—topsoil and subsoil.

Soil horizons are the layers in the soil as you move down the soil profile. A soil profile may have soil horizons that are easy or difficult to distinguish.

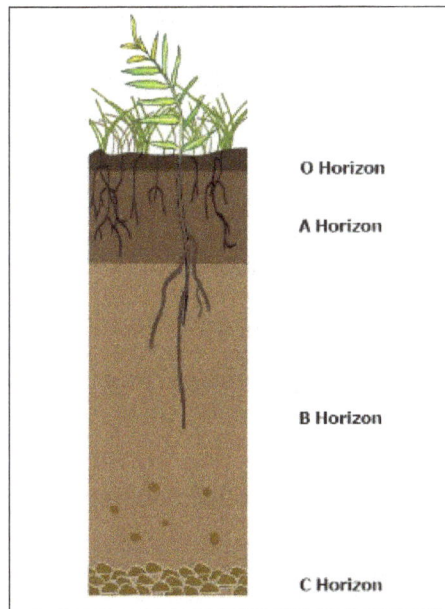

Soil profile showing the different layers or horizons

Most soils exhibit 3 main horizons:

- A horizon—humus-rich topsoil where nutrient, organic matter and biological activity are highest (i.e. most plant roots, earthworms, insects and micro-organisms are active). The A horizon is usually darker than other horizons because of the organic materials.

- B horizon—clay-rich subsoil. This horizon is often less fertile than the topsoil but holds more moisture. It generally has a lighter color and less biological activity than the A horizon. Texture may be heavier than the A horizon too.

- C horizon—underlying weathered rock (from which the A and B horizons form).

Some soils also have an O horizon mainly consisting of plant litter which has accumulated on the soil surface.

The properties of horizons are used to distinguish between soils and determine land-use potential.

Factors affecting Soil Formation

Soil forms continuously, but slowly, from the gradual breakdown of rocks through weathering. Weathering can be a physical, chemical or biological process:

- Physical weathering—breakdown of rocks from the result of a mechanical action. Temperature changes, abrasion (when rocks collide with each other) or frost can all cause rocks to break down.

- Chemical weathering—breakdown of rocks through a change in their chemical makeup. This can happen when the minerals within rocks react with water, air or other chemicals.

- Biological weathering—the breakdown of rocks by living things. Burrowing animals help water and air get into rock, and plant roots can grow into cracks in the rock, making it split.

The accumulation of material through the action of water, wind and gravity also contributes to soil formation. These processes can be very slow, taking many tens of thousands of years. Five main interacting factors affect the formation of soil:

- Parent material—minerals forming the basis of soil.

- Living organisms—influencing soil formation.

- Climate—affecting the rate of weathering and organic decomposition.

- Topography—grade of slope affecting drainage, erosion and deposition.

- Time—influencing soil properties.

Interactions between these factors produce an infinite variety of soils across the earth's surface.

Parent Materials

Soil minerals form the basis of soil. They are produced from rocks (parent material) through the processes of weathering and natural erosion. Water, wind, temperature change, gravity, chemical interaction, living organisms and pressure differences all help break down parent material.

The types of parent materials and the conditions under which they break down will influence the properties of the soil formed. For example, soils formed from granite are often sandy and infertile whereas basalt under moist conditions breaks down to form fertile, clay soils.

Organisms

Soil formation is influenced by organisms (such as plants), micro-organisms (such as bacteria or fungi), burrowing insects, animals and humans.

As soil forms, plants begin to grow in it. The plants mature, die and new ones take their place. Their leaves and roots are added to the soil. Animals eat plants and their wastes and eventually their bodies are added to the soil.

This begins to change the soil. Bacteria, fungi, worms and other burrowers break down plant litter and animal wastes and remains, to eventually become organic matter. This may take the form of peat, humus or charcoal.

Climate

Temperature affects the rate of weathering and organic decomposition. With a colder and drier climate, these processes can be slow but, with heat and moisture, they are relatively rapid.

Rainfall dissolves some of the soil materials and holds others in suspension. The water carries or leaches these materials down through the soil. Over time this process can change the soil, making it less fertile.

Topography

The shape, length and grade of a slope affect drainage. The aspect of a slope determines the type of vegetation and indicates the amount of rainfall received. These factors change the way soils form.

Soil materials are progressively moved within the natural landscape by the action of water, gravity and wind (for example, heavy rains erode soils from the hills to lower areas, forming deep soils). The soils left on steep hills are usually shallower. Transported soils include:

- Alluvial (water transported)
- Colluvial (gravity transported)
- Aeolian (wind transported) soils

Time

Soil properties may vary depending on how long the soil has been weathered.

Minerals from rocks are further weathered to form materials such as clays and oxides of iron and aluminum.

References
- Geomorphic-processes, geography: exampariksha.com, Retrieved 11 March, 2019
- Mass-movement, science: britannica.com, Retrieved 13 June, 2019
- Fundamentals: physicalgeography.net, Retrieved 10 January, 2019
- Forms, soil, management, land, environment: qld.gov.au, Retrieved 26 February, 2019

Chapter 5
Landforms and their Evolution

The small to medium tracts of Earth's surface are called landforms. The evolution of landforms occurs due to interaction between different physical processes and environmental factors, such as tectonics, underlying rock structures, climate and climatic changes, rock types, and human activities. This chapter discusses in detail these theories and methodologies related to landforms and their evolution.

The present day forms of land surface (landform) are a result of different earth surface processes that operated over long geological times, landform is usually the first and easiest thing we observe when we study global change and the impacts of human activities on our environment and may contain important clues to past processes related to global change and human impacts. In order to be able to improve and maintain the sustainability of our environment and predict and reduce the impact of contemporary earth surface processes that lead to natural hazards (such as landslides), we need to have a basic understanding of the general configuration of landforms and of the surface processes and environmental factors involved in their formation and evolution. Landform evolution is an important aspect of earth sciences and involves complicated interaction among different physical processes and environmental factors, such as underlying rock structures, tectonics, rock types, climate and climatic changes, and human activities, all occurring over a wide range of spatial and temporal scales. However, because of the degree of complexity in spatial and temporal scales, long-term landform evolution cannot be observed directly. Further, the interacting processes involved are hard to infer from the limited temporal observations of present day forms.

Factors influencing Landform Evolution

One of the most commonly observed patterns of river systems is the branching pattern of dendritic drainage network (from the Greek dendrites). It ranges from the small scale of rill (formed by erosion on a newly exposed surface) to the continental scale drainages that evolved over long geological times (e.g., the Mississippi, the Amazon, the Congo, and the Yellow). The branching drainage networks get started and controls their evolution.

Dendritic drainage pattern on the Western Plains.

Dendritic drainage pattern in Yemen.

Overland flow (runoff) of waters with erosive potential is generated when the volume of water supplied from rainfall or snowmelt exceeds the infiltration capacity of the soils or substrate. For slopes developing in arid/semi-arid regions (sparse protective vegetation cover), rain splash erosion can be very effective. Infrequent but substantial rainfall events may be sufficient to generate runoff in which the shear stress of overland flow exceeds the shear strength of the surficial materials such that weathered or loose particles are eroded and transported down slope. Spatial variation in surface topography of the slope and texture of the slope sediments often leads to gulling for landscapes where low vegetation cover and root mass are insufficient to stabilize the surficial materials. These are sometimes referred to as wash dominated slopes in which the rate of fluvial erosion, although infrequent, exceeds the rate of weathering which supplies the erodible sediment.

Climatic variables play a key role in drainage form, slope form and process, and in the evolution of a drainage basin through time. Annual variations in temperature, precipitation, and seasonality of precipitation work together to influence the degree of chemical and physical weathering of slope materials, the depth of weathered materials or soils that develop, and perhaps most importantly, to determine the vegetation type and percentage of cover across a landscape. Vegetation covers in turn controls slope form and mass movement process and therefore the resultant drainage basin attributes.

In temperate regions, the rates of chemical and physical weathering are sufficiently high to produce thicker sequences of weathered materials or soils that often bury rock outcrops in their own weathering products. Vegetation cover is high, protecting the surface from rain splash, and the root mass is sufficient to stabilize the materials on the slope. When overland flow does occur it is often ineffective at eroding the surface because of the protective vegetation, and infiltrating waters moving downslope as "through flow" (water moving through permeable soil horizons) are prevented from eroding the soil because of the binding affects of plant roots. However, "piping" of waters flowing through small conduits (mm/cm scale) developed in permeable soil horizons can exert sufficient shear stress or fluid drag on soil particles of the pipe wall to transport them downslope. In the temperate climate landscape, downslope movement of materials to the fluvial channel occurs primarily by the slow mass movement process of either continuous or seasonal creep. Longitudinal profiles for these "creep dominated slopes" assume a smooth convex/concave form from the drainage divide downward to the stream channel.

Slope change from steep to gentle in the land surface could lead to the reduction of stream carrying capacity and accumulation of deposition at the base of the slope, usually in the form of alluvial fans or coalesced alluvial fans. The slope change could be caused by tectonic uplift, faulting, or differential erosion between two types of rock with different erodibility.

Running Water

Running water and groundwater are geomorphic agents which cause erosion and deposition. They form various erosional (destructional) and depositional (constructional) landforms. It is important to consider that these erosional and depositional activities and their landform creation are always aided by weathering and mass movements. There are some other independent controls like (i) stability of sea level; (ii) tectonic stability of landmass; (iii) climate etc. which influence the evolution of these landforms.

Benefits of Running Water

- Running water, which doesn't need any further explanation, has two components: one is overland flow on the general land surface as a sheet and the other is linear flow as streams and rivers in valleys.

- The overland flow causes sheet erosion and depending upon the irregularities of the land surface, the overland flow may concentrate into narrow to wide paths.

- During the sheet erosion, minor or major quantities of materials from the surface of the land are removed in the direction of flow and gradual small and narrow rills will form.

- These rills will gradually develop into long and wide gullies, the gullies will further deepen, widen and lengthen and unite to give rise to a network of valleys.

- Once a valley is formed, it later develops into a stream or river.

Rills → Gullies → Valley

Courses of a River

A river, which is the best example of the linear flow of running water through a valley, can be divided into three, on the basis of its course – upper course, middle course and lower course.

Upper Course/Stage of Youth (Erosion Dominates)

- It starts from the source of the river in hilly or mountainous areas.

- The river flows down the steep slope and, as a result, its velocity and eroding power are at their maximum.

- Streams are few, with poor integration.

- As the river flows down with high velocity, vertical erosion or downward cutting will be high which results in the formation of V-Shaped Valleys.

- Waterfalls, rapids, and gorges exist where the local hard rock bodies are exposed.

Middle Course/Stage of Maturity (Transportation Dominates)

- In this stage, vertical erosion slowly starts to replace with lateral erosion or erosion from both sides of the channel.

- Thus, the river channel causes the gradual disappearance of its V-shaped valley (not completely).

- Streams are plenty at this stage with good integration.

- Wider flood plains start to visible in this course and the volume of water increases with the confluence of many tributaries.

- The work of river predominantly becomes transportation of the eroded materials from the upper course (little deposition too).

- Landforms like alluvial fans, piedmont alluvial plains, meanders etc. can be seen at this stage.

Lower Course/Stage of Old (Deposition Dominates)

- The river starts to flow through a broad, level plain with heavy debris brought down from upper and middle courses.

- Vertical erosion has almost stopped and lateral erosion still goes on.

- The work of the river is mainly deposition, building up its bed and forming an extensive flood plain.

- Landforms like braided channels, floodplains, levees, meanders, oxbow lakes, deltas etc. can be seen at this stage.

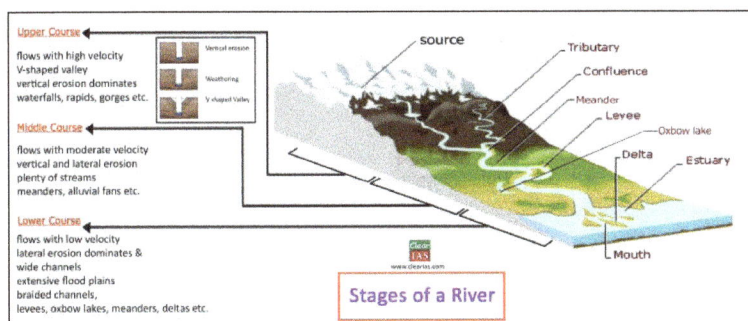

Stages of a River

Running Water: Erosion, Transportation and Deposition

- Erosion occurs when overland flow moves soil particles downslope.

- The rock materials carried by erosion is the load of the river.

- This load acts as a grinding tool helping in cutting the bottom and sides of the river bed, resulting in deepening and widening of the river channel.

Erosion Types

The work of river erosion is accomplished in different ways, all of which may operate together. They are corrasion, corrosion, hydraulic action etc.

1. Corrasion or Abration: As the rock particles bounce, scrape and drag along the bottom and sides of the river, they break off additional rock fragments. This form of erosion is called corrasion or abration. They are two types: vertical corrosion which acts downward and lateral corrosion which acts on both sides.

2. Corrosion or Solution: This is the chemical or solvent action of water on soluble or partly soluble rocks with which the river water comes in contact.

3. Hydraulic Action: This is the mechanical loosening and sweeping away of material by the sheer force or river water itself. No load or material is involved in this process.

Transportation Types

After erosion, the eroded materials get transported with the running water. This transportation of eroded materials is carried in four ways:

1. Traction: The heavier and larger rock fragments like gravels, pebbles etc. are forced by the flow of the river to roll along its bed. These fragments can be seen rolling, slipping, bumping and being dragged. This process is called as traction and the load transported in this way are called traction load.

2. Saltation: Some of the fragments of the rocks move along the bed of a stream by jumping or bouncing continuously. This process is called as saltation.

3. Suspension: The holding up of small particles of sand, silt and mud by the water as the stream flows is called suspension.

4. Solution: Some parts of the rock fragments dissolved in the river water and transported. This type of transportation is called solution transportation.

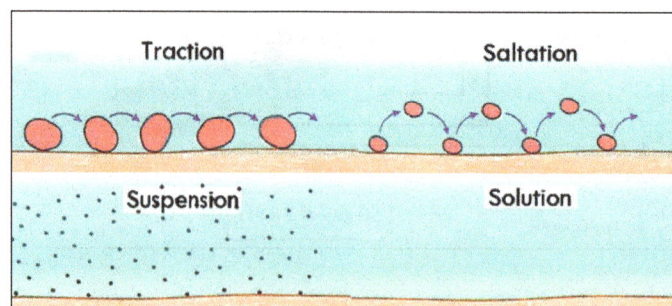

- When the stream comes down from the hills to plain areas with the eroded and transported materials, the absence of slope/gradient causes the river to lose it energy to further carry those transported materials.

- As a result, the load of the river starts to settle down which is termed as deposition.

- Erosion, transportation, and deposition continue until the slopes are almost completely flattened leaving finally a lowland of faint relief called peneplains with some low resistant remnants called monadnocks.

Erosional Landforms due to Running Water

Valleys, Gorges and Canyon

- Valleys are formed as a result of running water.

- The rills which are formed by the overland flow of water later develop into gullies.

- These gullies gradually deepen and widen to form valleys.

- A gorge is a deep valley with very steep to straight sides.

- A canyon is characterized by steep step-like side slopes and may be as deep as a gorge.

- A gorge is almost equal in width at its top as well as bottom and is formed in hard rocks while a canyon is wider at its top than at its bottom and is formed in horizontal bedded sedimentary rocks.

Gorge   Canyon

Potholes and Plunge Pools

- Potholes are more or less circular depressions over the rocky beds of hills streams.

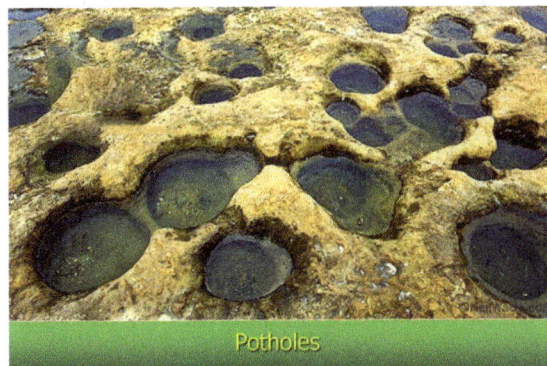
Potholes

- Once a small and shallow depression forms, pebbles and boulders get collected in those depressions and get rotated by flowing water. Consequently, the depressions grow in dimensions to form potholes.

- Plunge pools are nothing but large, deep potholes commonly found at the foot of a waterfall.

- They are formed because of the sheer impact of water and rotation of boulders.

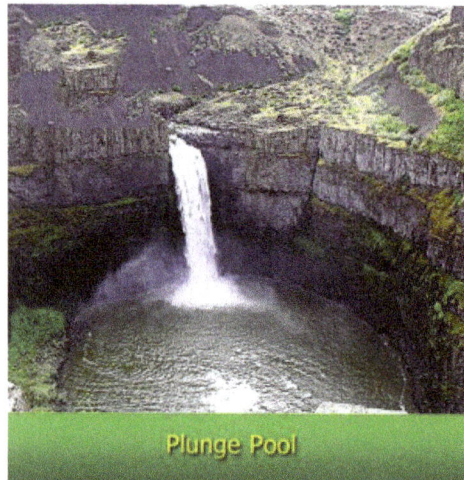
Plunge Pool

Incised or Entrenched Meanders

- They are very deep wide meanders (loop-like channels) found cut in hard rocks.

- In the course of time, they deepen and widen to form gorges or canyons in hard rock.

- The difference between a normal meander and an incised/entrenched meander is that the latter found on hard rocks.

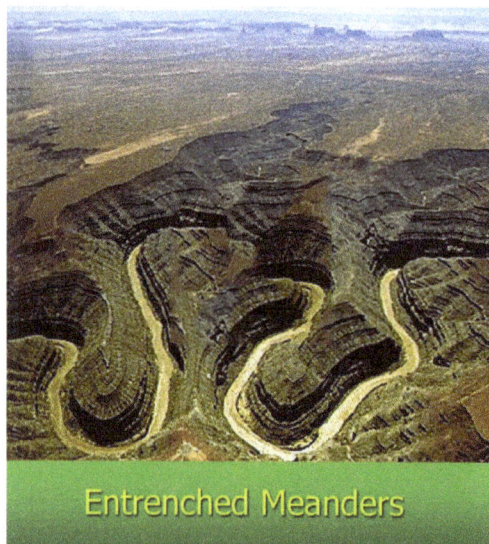
Entrenched Meanders

River Terraces

- They are surfaces marking old valley floor or flood plains.

- They are basically the result of vertical erosion by the stream.

- When the terraces are of the same elevation on either side of the river, they are called as paired terraces.

- When the terraces are seen only on one side with none on the other or one at quite a different elevation on the other side, they are called as unpaired terraces.

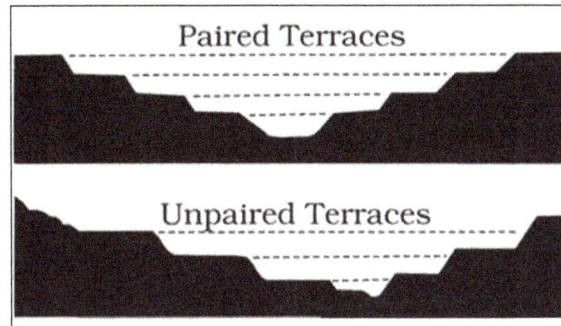

Depositional Landforms due to Running Water

Alluvial Fans

- They are found in the middle course of a river at the foot of slope/mountains.

- When the stream moves from the higher level break into foot slope plain of low gradient, it loses its energy needed to transport much of its load.

- Thus, they get dumped and spread as a broad low to the high cone-shaped deposits called an alluvial fan.

- The deposits are not roughly very well sorted.

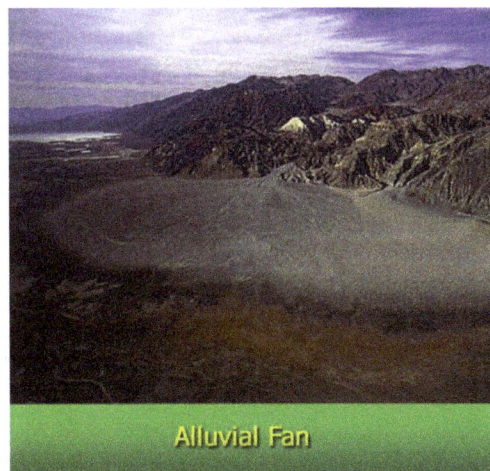

Deltas

- Deltas are like an alluvial fan but develop at a different location.

- They are found in the mouth of the river, which is the final location of depositional activity of a river.

- Unlike alluvial fans, the deposits making up deltas are very well sorted with clear stratification.

- The coarser material settle out first and the finer materials like silt and clay are carried out into the sea.

Flood Plains and Natural Levees

- Deposition develops a flood plain just as erosion makes valleys.

- A riverbed made of river deposits is the active flood plain and the flood plain above the bank of the river is the inactive flood plain.

- Natural levees are found along the banks of large rivers. They are low, linear and parallel ridges of coarse deposits along the banks of a river.

- The levee deposits are coarser than the deposits spread by flood water away from the river.

Meanders and Oxbow Lakes

- Meanders are loop-like channel patterns develop over the flood and delta plains.

- They are actually not a landform but only a type of channel pattern formed as a result of deposition.

- They are formed basically because of three reasons: (i) propensity of water flowing over very gentle gradient to work laterally on the banks; (ii) unconsolidated nature of alluvial deposits making up the bank with many irregularities; (iii) Coriolis force acting on fluid water deflecting it like deflecting the wind.

- The concave bank of a meander is known as cut-off bank and the convex bank is known as a slip-off

- As meanders grow into deep loops, the same may get cut-off due to erosion at the inflection point and are left as oxbow lakes.

- For large rivers, the sediments deposited in a linear fashion at the depositional side of a meander are called as Point Bars or Meander Bars.

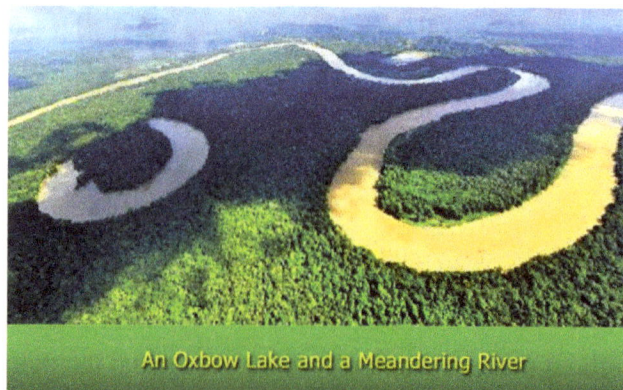

An Oxbow Lake and a Meandering River

Braided Channels

Braided Channels

- When selective deposition of coarser materials causes the formation of a central bar, it diverts the flow of river towards the banks, which increases lateral erosion.

- Similarly, when more and more such central bars are formed, braided channels are formed.

- Riverine Islands are the result of braided channels.

Benefits of Groundwater

- The part of rain or snow-melt water which accumulates in the rocks after seeping through the surface is called underground water or simply groundwater.

- The rocks through which water can pass easily are called as permeable rocks while the rocks which do not allow water to pass are called as impermeable rocks.

- After vertically going down to some depth, the water under the ground flows horizontally through the bedding planes, joints or through the materials themselves.

- Although the amount of groundwater varies from place to place, its role in shaping the surface features of the earth is quite important.

- The works of groundwater are mainly seen in rocks like limestone, gypsum or dolomite which are rich in calcium carbonate.

- Any limestone, dolomite or gypsum region showing typical landforms produced by the action of groundwater through the process of solution and deposition is called as Karst Topography (Karst region in the Balkans).

- The zones or horizons of permeable and porous rocks which are fully filled with water are called as the Zones of Saturation.

- The marks which show the upper surface of these saturated zones of the groundwater are called as the Water Tables.

- And these rocks, which are filled with underground water, are called as aquifers.

- The water table is generally higher in the areas of high precipitation and also in areas bordering rivers and lakes.

- They also vary according to seasons. On the basis of variability, water tables are of two types: (i) Permanent water table, in which the water will never fall below a certain level and wells dug up to this depth provide water in all seasons; (ii) Temporary water tables, which are seasonal water tables.

- Springs: They are the surface outflow of groundwater through an opening in a rock under hydraulic pressure.

- When such springs emit hot water, they are called as Hot Springs. They generally occur in areas of active or recent volcanism.

- When a spring emits hot water and steam in the form of fountains or jets at regular intervals, they are called as geysers.

- In a geyser, the period between two emissions is sometimes regular (Yellowstone National Park of USA is the best example).

Erosional Landforms due to Groundwater

Sinkholes and caves are erosional landforms formed due to the action of ground water.

Sinkholes

- Small to medium sized rounded to sub-rounded shallow depressions called swallow holes forms on the surface of rocks like limestone by the action of the solution.

- A sinkhole is an opening more or less circular at the top and funnel-shaped towards the bottom.

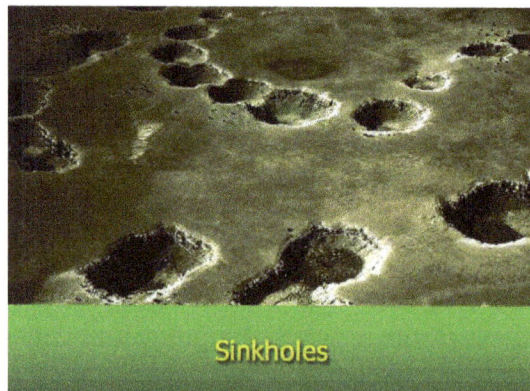

Sinkholes

- When as sinkhole is formed solely through the process of solution, it is called as a solution sink.

- Some sinkhole starts its formation through the solution process but later collapse due to the presence of some caves or hollow beneath it and becomes a bigger sinkhole. These types are called as collapse sinks.

- The term Doline is sometimes used to refer collapse sinks.

- Solution sinks are more common than collapse sinks.

- When several sink holes join together to form valley of sinks, they are called as valley sinks or Uvalas.

- Lapies are the irregular grooves and ridges formed when most of the surfaces of limestone are eaten by solution process.

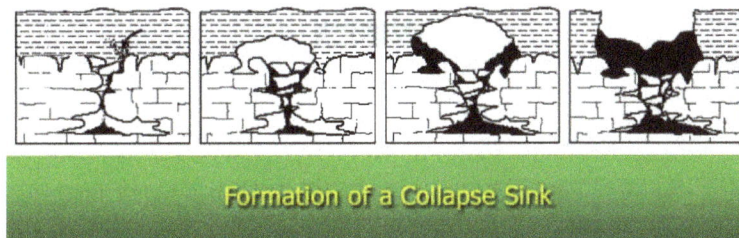

Formation of a Collapse Sink

Caves

- In the areas where there are alternative beds of rocks (non-soluble) with limestone or dolomite in between or in areas where limestone are dense, massive and occurring as thick beds, cave formation is prominent.

- Caves normally have an opening through which cave streams are discharged.

- Caves having an opening at both the ends are called tunnels.

- Depositional Landforms of Groundwater.

- Stalactites and stalagmites.

- They are formed when the calcium carbonates dissolved in groundwater get deposited once the water evaporates.

- These structures are commonly found in limestone caves.

- Stalactites are calcium carbonate deposits hanging as icicles while Stalagmites are calcium carbonate deposits which rise up from the floor.

- When a stalactite and stalagmite happened to join together, it gives rise to pillars or columns of different diameters.

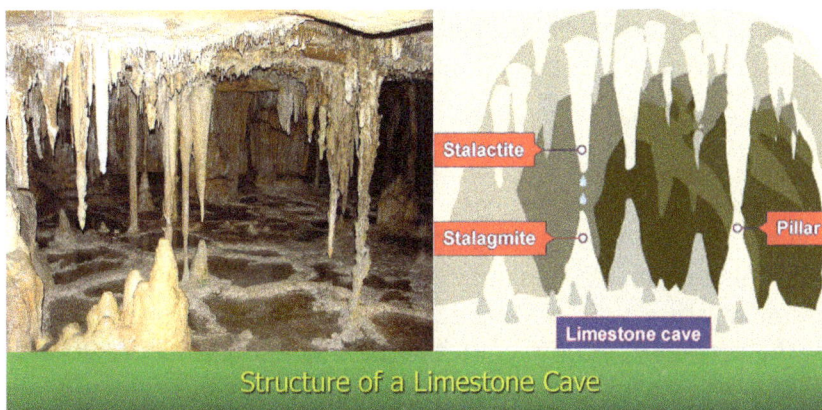

Structure of a Limestone Cave

Glaciers

Glacial landform, are products of flowing ice Glacial landform, any product of flowing ice and melt water. Such landforms are being produced today in glaciated areas, such as Greenland, Antarctica, and many of the world's higher mountain ranges. In addition, large expansions of present-day glaciers have recurred during the course of Earth history. At the maximum of the last ice age, which ended about 20,000 to 15,000 years ago, more than 30 percent of the Earth's land surface was covered by ice. Consequently, if they have not been obliterated by other landscape-modifying processes since that time, glacial landforms may still exist in regions that were once glaciated but are now devoid of glaciers.

Periglacial features, which form independently of glaciers, are nonetheless a product of the same cold climate that favors the development of glaciers, and so are treated here as well.

General Considerations

Before describing the different landforms produced by glaciers and their meltwater, the glacial environment and the processes responsible for the formation of such landforms is briefly

Types of Glaciers

There are numerous types of glaciers, but it is sufficient here to focus on two broad classes: mountain, or valley, glaciers and continental glaciers, or ice sheets, (including ice caps).

Generally, ice sheets are larger than valley glaciers. The main difference between the two classes, however, is their relationship to the underlying topography. Valley glaciers are rivers of ice usually found in mountainous regions, and their flow patterns are controlled by the high relief in those areas. In map view, many large valley glacier systems, which have numerous tributary glaciers that join to form a large "trunk glacier," resemble the roots of a plant. Pancake like ice sheets, on the other hand, are continuous over extensive areas and completely bury the underlying landscape beneath hundreds or thousands of metres of ice. Within continental ice sheets, the flow is directed more or less from the centre outward. At the periphery, however, where ice sheets are much thinner, they may be controlled by any substantial relief existing in the area. In this case, their borders may be lobate on a scale of a few kilometres, with tonguelike protrusions called outlet glaciers. Viewed by themselves, these are nearly indistinguishable from the lower reaches of a large valley glacier system. Consequently, many of the landforms produced by valley glaciers and continental ice sheets are similar or virtually identical, though they often differ in magnitude.

Glacial Erosion

Two processes, internal deformation and basal sliding, are responsible for the movement of glaciers under the influence of gravity. The temperature of glacier ice is a critical condition that affects these processes. For this reason, glaciers are classified into two main types, temperate and polar, according to their temperature regime. Temperate glaciers are also called isothermal glaciers, because they exist at the pressure-melting point (the melting temperature of ice at a given pressure) throughout their mass. The ice in polar or cold glaciers, in contrast, is below the pressure-melting point. Some glaciers have an intermediate thermal character. For example, subpolar glaciers are temperate in their interior parts, but their margins are cold-based. This classification is a broad generalization, however, because the thermal condition of a glacier may show wide variations in both space and time.

Internal deformation, or strain, in glacier ice is a response to shear stresses arising from the weight of the ice (ice thickness) and the degree of slope of the glacier surface. Internal deformation occurs by movement within and between individual ice crystals (slow creep) and by brittle failure (fracture), which arises when the mass of ice cannot adjust its shape rapidly enough by the creep process to take up the stresses affecting it. The relative importance of these two processes is greatly influenced by the temperature of the ice. Thus, fractures due to brittle failure under tension, known as crevasses, are usually much deeper in polar ice than they are in temperate ice.

The temperature of the basal ice is an important influence upon a glacier's ability to erode its bed. When basal temperatures are below the pressure-melting point, the ability of the ice mass to slide on the bed (basal sliding) is inhibited by the adhesion of the basal ice to the frozen bed beneath. Basal sliding is also diminished by the greater rigidity of polar ice: this reduces the rate of creep, which, in turn, reduces the ability of the more rigid ice to deform around obstacles on the glacier bed. Thus, the flow of cold-based glaciers is predominantly controlled by internal deformation, with proportionately low rates of basal sliding. For this reason, rates of abrasion are commonly low beneath polar glaciers, and slow rates of erosion commonly result. Equally, the volume of meltwater is frequently very low, so that the extent of sediments and landforms derived from polar glaciers is limited.

Temperate glaciers, being at the pressure-meeting point, move by both mechanisms, with basal sliding being the more important. It is this sliding that enables temperate glaciers to erode their beds and carve landforms so effectively. Ice is, however, much softer and has a much lower shear strength than most rocks, and pure ice alone is not capable of substantially eroding anything other than unconsolidated sediments. Most temperate glaciers have a basal debris zone from several centimetres to a few metres thick that contains varying amounts of rock debris in transit. In this respect, glaciers act rather like sheets of sandpaper; while the paper itself is too soft to sand wood, the adherent hard grains make it a powerful abrasive system. The analogy ends here, however, for the rock debris found in glaciers is of widely varying sizes—from the finest rock particles to large boulders—and also generally of varied types as it includes the different rocks that a glacier is overriding. For this reason, a glacially abraded surface usually bears many different "tool-marks," from microscopic scratches to gouges centimetres deep and tens of metres long. Over thousands of years glaciers may erode their substrate to a depth of several tens of metres by this mechanism, producing a variety of streamlined landforms typical of glaciated landscapes.

Several other processes of glacial erosion are generally included under the terms glacial plucking or quarrying. This process involves the removal of larger pieces of rock from the glacier bed. Various explanations for this phenomenon have been proposed. Some of the mechanisms suggested are based on differential stresses in the rock caused by ice being forced to flow around bedrock obstacles. High stress gradients are particularly important, and the resultant tensile stresses can pull the rock apart along pre-existing joints or crack systems. These pressures have been shown to be sufficient to fracture solid rock, thus making it available for removal by the ice flowing above it. Other possibilities include the forcing apart of rock by the pressure of crystallization produced beneath the glacier as water derived from the ice refreezes (regelation) or because of temperature fluctuations in cavities under the glacier. Still another possible mechanism involves hydraulic pressures of flowing water known to be present, at least temporarily, under nearly all temperate glaciers. It is hard to determine which process is dominant because access to the base of active glaciers is rarely possible. Nonetheless, investigators know that larger pieces of rock are plucked from the glacier bed and contribute to the number of abrasive "tools" available to the glacier at its base. Other sources for the rock debris in glacier ice may include rockfalls from steep slopes bordering a glacier or unconsolidated sediments overridden as a glacier advances.

Glacial Deposition

Debris in the glacial environment may be deposited directly by the ice (till) or, after reworking, by meltwater streams (outwash). The resulting deposits are termed glacial drift.

As the ice in a valley glacier moves from the area of accumulation to that of ablation, it acts like a conveyor belt, transporting debris located beneath, within, and above the glacier toward its terminus or, in the case of an ice sheet, toward the outer margin. Near the glacier margin where the ice velocity decreases greatly is the zone of deposition. As the ice melts away, the debris that was originally frozen into the ice commonly forms a rocky and muddy blanket over the glacier margin. This layer often slides off the ice in the form of mudflows. The resulting deposit is called a flow-till by some authors. On the other hand, the debris may be laid down more or less in place as the ice melts away around and beneath it. Such deposits are referred to as melt-out till, and sometimes as ablation till. In many cases, the material located between a moving glacier and its bedrock bed is severely sheared, compressed, and "over-compacted." This type of deposit is called lodgment till. By definition, till is any material laid down directly or reworked by a glacier. Typically, it is a mixture of rock fragments and boulders in a fine-grained sandy or muddy matrix (non-stratified drift). The exact composition of any particular till, however, depends on the materials available to the glacier at the time of deposition. Thus, some tills are made entirely of lake clays deformed by an overriding glacier. Other tills are composed of river gravels and sands that have been "bulldozed" and striated during a glacial advance. Tills often contain some of the tools that glaciers use to abrade their bed. These rocks and boulders bear striations, grooves, and facets, and characteristic till-stones are commonly shaped like bullets or flat-irons. Till-boulders of a rock type different from the bedrock on which they are deposited are dubbed "erratics." In some cases, erratics with distinctive lithologies can be traced back to their source, enabling investigators to ascertain the direction of ice movement of ice sheets in areas where striations either are absent or are covered by till or vegetation.

Meltwater deposits, also called glacial outwash, are formed in channels directly beneath the glacier or in lakes and streams in front of its margin. In contrast to till, outwash is generally bedded or laminated (stratified drift), and the individual layers are relatively well sorted according to grain size. In most cases, gravels and boulders in outwash are rounded and do not bear striations or grooves on their surfaces, since these tend to wear off rapidly during stream transport. The grain size of individual deposits depends not only on the availability of different sizes of debris but also on the velocity of the depositing current and the distance from the head of the stream. Larger boulders are deposited by rapidly flowing creeks and rivers close to the glacier margin. Grain size of deposited material decreases with increasing distance from the glacier. The finest fractions, such as clay and silt, may be deposited in glacial lakes or ponds or transported all the way to the ocean.

Finally, it must be stressed that most glacier margins are constantly changing chaotic masses of ice, water, mud, and rocks. Ice-marginal deposits thus are of a highly variable nature over short distances, as is much the case with till and outwash as well.

Erosional Landforms

Small-scale Features of Glacial Erosion

Glacial erosion is caused by two different processes: abrasion and plucking. Nearly all glacially scoured erosional landforms bear the tool-marks of glacial abrasion provided that they have not been removed by subsequent weathering. Even though these marks are not large enough to be called landforms, they constitute an integral part of any glacial landscape.The type of mark produced on a surface during glacial erosion depends on the size and shape of the tool, the pressure being applied to it, and the relative harnesses of the tool and the substrate.

Rock Polish

The finest abrasive available to a glacier is the so-called rock flour produced by the constant grinding at the base of the ice. Rock flour acts like jewelers' rouge and produces microscopic scratches, which with time smooth and polish rock surfaces, often to a high luster.

Striations

These are scratches visible to the naked eye, ranging in size from fractions of a millimetre to a few millimetres deep and a few millimetres to centimetres long. Large striations produced by a single tool may be several centimetres deep and wide and tens of metres long.

Because the striation-cutting tool was dragged across the rock surface by the ice, the long axis of a striation indicates the direction of ice movement in the immediate vicinity of that striation. Determination of the regional direction of movement of former ice sheets, however, requires measuring hundreds of striation directions over an extended area because ice moving close to the base of a glacier is often locally deflected by bedrock obstacles. Even when such a regional study is conducted, additional information is frequently needed in low-relief areas to determine which end of the striations points down-ice toward the former outer margin of the glacier. On an outcrop scale, such information can be gathered by studying "chatter marks." These crescentic gouges and lunate fractures are caused by the glacier dragging a rock or boulder over a hard and brittle rock surface and forming a series of sickle-shaped gouges. Such depressions in the bedrock are steep-sided on their "up-glacier" face and have a lower slope on their down-ice side. Depending on whether the horns of the sickles point up the glacier or down it, the chatter marks are designated crescentic gouges or lunate fractures. Another small-scale feature that allows absolute determination of the direction in which the ice moved is what is termed knob-and-tail. A knob-and-tail is formed during glacial abrasion of rocks that locally contain spots more resistant than the surrounding rock, as is the case, for example, with silicified fossils in limestone. After abrasion has been active for some time, the harder parts of the rock form protruding knobs as the softer rock is preferentially eroded away around them. During further erosion, these protrusions protect the softer rock on their lee side and a tail forms there, pointing from the knob to the margin of the glacier. The scale of these features depends primarily on the size of the in homogeneities in the rock and ranges from fractions of millimetres to metres.

P-forms and Glacial Grooves

These features, which extend several to tens of metres in length, are of uncertain origin. P-forms (P for plastically molded) are smooth-walled, linear depressions which may be straight, curved, or sometimes hairpin-shaped and measure tens of centimetres to metres in width and depth. Their cross sections are often semicircular to parabolic, and their walls are commonly striated parallel to their long axis, indicating that ice once flowed in them. Straight P-forms are frequently called glacial grooves, even though the term is also applied to large striations, which, unlike the P-forms, were cut by a single tool. Some researchers believe that P-forms were not carved directly by the ice but rather were eroded by pressurized mud slurries flowing beneath the glacier.

Erosional Landforms of Valley Glaciers

Many of the world's higher mountain ranges—e.g., the Alps, the North and South American Cordilleras, the Himalayas, and the Southern Alps in New Zealand, as well as the mountains of Norway, including those of Spitsbergen—are partly glaciated today. During periods of the Pleistocene, such glaciers were greatly enlarged and filled most of the valleys with ice, even reaching far beyond the mountain front in certain places. Most scenic alpine landscapes featuring sharp mountain peaks, steep-sided valleys, and innumerable lakes and waterfalls are a product of several periods of glaciation.

Erosion is generally greater than deposition in the upper reaches of a valley glacier, whereas deposition exceeds erosion closer to the terminus. Accordingly, erosional landforms dominate the landscape in the high areas of glaciated mountain ranges.

Cirques, Tarns, U-Shaped Valleys, Arêtes and Horns

The heads of most glacial valleys are occupied by one or several cirques (or corries). A cirque is an amphitheatre-shaped hollow with the open end facing down-valley. The back is formed by an arcuate cliff called the headwall. In an ideal cirque, the headwall is semicircular in plan view. This situation, however, is generally found only in cirques cut into flat plateaus. More common are headwalls angular in map view due to irregularities in height along their perimeter. The bottom of many cirques is a shallow basin, which may contain a lake. This basin and the base of the adjoining headwall usually show signs of extensive glacial abrasion and plucking. Even though the exact process of cirque formation is not entirely understood, it seems that the part of the headwall above the glacier retreats by frost shattering and ice wedging. The rock debris then falls either onto the surface of the glacier or into the randkluft or bergschrund. Both names describe the crevasse between the ice at the head of the glacier and the cirque headwall. The rocks on the surface of the glacier are successively buried by snow and incorporated into the ice of the glacier. Because of a downward velocity component in the ice in the accumulation zone, the rocks are eventually moved to the base of the glacier. At that point, these rocks, in addition to the rock debris from the bergschrund, become the tools with which the glacier erodes, striates, and polishes the base of the headwall and the bottom of the cirque.

During the initial growth and final retreat of a valley glacier, the ice often does not extend beyond the cirque. Such a cirque glacier is probably the main cause for the formation of the basin scoured into the bedrock bottom of many cirques. Sometimes these basins are "over-deepened" several tens of metres and contain lakes called tarns.

In contrast to the situation in a stream valley, all debris falling or sliding off the sides and the head-walls of a glaciated valley is immediately removed by the flowing ice. Moreover, glaciers are generally in contact with a much larger percentage of a valley's cross section than equivalent rivers or creeks. Thus glaciers tend to erode the bases of the valley walls to a much greater extent than do streams, whereas a stream erodes an extremely narrow line along the lowest part of a valley. The slope of the adjacent valley walls depends on the stability of the bedrock and the angle of repose of the weathered rock debris accumulating at the base of and on the valley walls. For this reason, rivers tend to form V-shaped valleys. Glaciers, which inherit V-shaped stream valleys, reshape them drastically by first removing all loose debris along the base of the valley walls and then preferentially eroding the bedrock along the base and lower sidewalls of the valley. In this way, glaciated valleys assume a characteristic parabolic or U-shaped cross profile, with relatively wide and flat bottoms and steep, even vertical sidewalls. By the same process, glaciers tend to narrow the bedrock divides between the upper reaches of neighbouring parallel valleys to jagged, knife-edge ridges known as arêtes. Arêtes also form between two cirques facing in opposite directions. The low spot, or saddle, in the arête between two cirques is called a col. A higher mountain often has three or more cirques arranged in a radial pattern on its flanks. Head ward erosion of these cirques finally leaves only a sharp peak flanked by nearly vertical headwall cliffs, which are separated by arêtes. Such glacially eroded mountains are termed horns, the most widely known of which is the Matterhorn in the Swiss Alps.

Hanging Valleys

Large valley glacier systems consist of numerous cirques and smaller valley glaciers that feed ice into a large trunk glacier. Because of its greater ice discharge, the trunk glacier has greater erosive capability in its middle and lower reaches than smaller tributary glaciers that join it there. The main valley is therefore eroded more rapidly than the side valleys. With time, the bottom of the main valley becomes lower than the elevation of the tributary valleys. When the ice has retreated, the tributary valleys are left joining the main valley at elevations substantially higher than its bottom. Tributary valleys with such unequal or discordant junctions are called hanging valleys. In extreme cases where a tributary joins the main valley high up in the steep part of the U-shaped trough wall, waterfalls may form after deglaciation, as in Yosemite and Yellowstone national parks in the western United States.

Paternoster Lakes

Some glacial valleys have an irregular, longitudinal bedrock profile, with alternating short, steep steps and longer, relatively flat portions. Even though attempts have been made to explain this feature in terms of some inherent characteristic of glacial flow, it seems more likely that differential erodibility of the underlying bedrock is the real cause of the phenomenon. Thus the steps are probably formed by harder or less fractured bedrock, whereas the flatter portions between the steps are underlain by softer or more fractured rocks. In some cases, these softer areas have been excavated by a glacier to form shallow bedrock basins. If several of these basins are occupied by lakes along one glacial trough in a pattern similar to beads on a string, they are called paternoster lakes by analogy with a string of rosary beads.

Roches Moutonnées

These structures are bedrock knobs or hills that have a gently inclined, glacially abraded, and streamlined stoss side (i.e., one that faces the direction from which the overriding glacier impinged) and a steep, glacially plucked lee side. They are generally found where jointing or fracturing in the bedrock allows the glacier to pluck the lee side of the obstacle. In plan view, their long axes are often, but not always, aligned with the general direction of ice movement.

Rock Drumlins

A feature similar to roches moutonnées, rock drumlins are bedrock knobs or hills completely streamlined, usually with steep stoss sides and gently sloping lee sides. Both roches moutonnées and rock drumlins range in length from several metres to several kilometres and in height from tens of centimetres to hundreds of metres. They are typical of both valley and continental glaciers. The larger ones, however, are restricted to areas of continental glaciation.

Erosional Landforms of Continental Glaciers

In contrast to valley glaciers, which form exclusively in areas of high altitude and relief, continental glaciers, including the great ice sheets of the past, occur in high and middle latitudes in both hemispheres, covering landscapes that range from high alpine mountains to low-lying areas with negligible relief. Therefore, the landforms produced by continental glaciers are more diverse and widespread. Yet, just like valley glaciers, they have an area where erosion is the dominant process and an area close to their margins where net deposition generally occurs. The capacity of a continental glacier to erode its substrate has been a subject of intense debate. All of the areas formerly covered by ice sheets show evidence of areally extensive glacial scouring. The average depth of glacial erosion during the Pleistocene probably did not exceed a few tens of metres, however. This is much less than the deepening of glacial valleys during mountain glaciation. One of the reasons for the apparent limited erosional capacity of continental ice sheets in areas of low relief may be the scarcity of tools available to them in these regions. Rocks cannot fall onto a continental ice sheet in the accumulation zone, because the entire landscape is buried. Thus, all tools must be quarried by the glacier from the underlying bedrock. With time, this task becomes increasingly difficult as bedrock obstacles are abraded and streamlined. Nonetheless, the figure for depth of glacial erosion during the Pleistocene cited above is an average value, and locally several hundreds of metres of bedrock were apparently removed by the great ice sheets. Such enhanced erosion seems concentrated at points where the glaciers flowed from hard, resistant bedrock onto softer rocks or where glacial flow was channelized into outlet glaciers.

As a continental glacier expands, it strips the underlying landscape of the soil and debris accumulated at the preglacial surface as a result of weathering. The freshly exposed harder bedrock is then eroded by abrasion and plucking. During this process, bedrock obstacles are shaped into streamlined "whaleback" forms, such as roches moutonnées and rock drumlins. The adjoining valleys are scoured into rock-floored basins with the tools plucked from the lee sides of roches moutonnées. The long axes of the hills and valleys are often preferentially oriented in the direction of ice flow. An area totally composed of smooth whaleback forms and basins is called a streamlined landscape.

Streams cannot erode deep basins because water cannot flow uphill. Glaciers, on the other hand, can flow uphill over obstacles at their base as long as there is a sufficient slope on the upper ice surface pointing in that particular direction. Therefore the great majority of the innumerable lake basins and small depressions in formerly glaciated areas can only be a result of glacial erosion. Many of these lakes, such as the Finger Lakes in the U.S. state of New York, are aligned parallel to the direction of regional ice flow. Other basins seem to be controlled by preglacial drainage systems. Yet, other depressions follow the structure of the bedrock, having been preferentially scoured out of areas underlain by softer or more fractured rock.

A number of the largest freshwater lake basins in the world (e.g., the Great Lakes or the Great Slave Lake and Great Bear Lake in Canada) are situated along the margins of the Precambrian shield of Many researchers believe that glacial erosion was especially effective at these locations because the glaciers could easily abrade the relatively soft sedimentary rocks to the south with hard, resistant crystalline rocks brought from the shield areas that lie to the north. Nonetheless, further research is necessary to determine how much of the deepening of these features can be ascribed to glacial erosion, as opposed to other processes such as tectonic activity or preglacial stream erosion.

Fjords are found along some steep, high-relief coast-lines where continental glaciers formerly flowed into the sea. They are deep, narrow valleys with U-shaped cross sections that often extend inland for tens or hundreds of kilometres and are now partially drowned by the ocean. These troughs are typical of the Norwegian coast, but they also are found in Canada, Alaska, Iceland, Greenland, Antarctica, New Zealand, and southernmost Chile. The floor and steep walls of fjords show ample evidence of glacial erosion. The long profile of many fjords, including alternating basins and steps, is very similar to that of glaciated valleys. Toward the mouth, fjords may reach great depths, as in the case of Sogn Fjord in southern Norway where the maximum water depth exceeds 1,300 metres. At the mouth of a fjord, however, the floor rises steeply to create a rock threshold, and water depths decrease markedly. At Sogn Fjord the water at this "threshold" is only 150 metres deep, and in many fjords the rock platform is covered by only a few metres of water. The exact origin of fjords is still a matter of debate. While some scientists favour a glacial origin, others believe that much of the relief of fjords is a result of tectonic activity and that glaciers only slightly modified preexisting large valleys. In order to erode Sogn Fjord to its present depth, the glacier occupying it during the maximum of the Pleistocene must have been 1,800 to 1,900 metres thick. Such an ice thickness may seem extreme, but even now, during an interglacial period, the Skelton Glacier in Antarctica has a maximum thickness of about 1,450 metres. This outlet glacier of the Antarctic ice sheet occupies a trough, which in places is more than one kilometre below sea level and would become a fjord in the event of a large glacial retreat.

Depositional Landforms of Valley Glaciers

Moraines

As a glacier moves along a valley, it picks up rock debris from the valley walls and floor, transporting it in, on, or under the ice. As this material reaches the lower parts of the glacier where ablation is dominant, it is concentrated along the glacier margins as more and more debris melts out of the ice. If the position of the glacier margin is constant for an extended amount of time, larger accumulations of glacial debris will form at the glacier margin. In addition, a great deal of material is rapidly flushed through and out of the glacier by meltwater streams flowing under, within, on, and next to the glacier. Part of this streamload is deposited in front of the glacier close to its snout. There, it may mix with material brought by, and melting out from, the glacier as well as with material washed in from other, nonglaciated tributary valleys. If the glacier then advances or readvances after a time of retreat, it will "bulldoze" all the loose material in front of it into a ridge of chaotic debris that closely hugs the shape of the glacier snout. Any such accumulation of till melted out directly from the glacier or piled into a ridge by the glacier is a moraine. Large valley glaciers are capable of forming moraines a few hundred metres high and many hundreds of metres wide. Linear accumulations of till formed immediately in front of or on the lower end of the glacier are end moraines. The moraines formed along the valley slopes next to the side margins of the glacier are termed lateral moraines. During a single glaciation, a glacier may form many such moraine arcs, but all the smaller moraines, which may have been produced during standstills or short advances while the glacier moved forward to its outermost ice position, are generally destroyed as the glacier resumes its advance. The end moraine of largest extent formed by the glacier (which may not be as extensive as the largest ice advance) during a given glaciation is called the terminal moraine of that glaciation. Successively smaller moraines formed during standstills or small readvances as the glacier retreats from the terminal moraine position are recessional moraines.

Flutes

The depositional equivalent of erosional knob-and-tail structure is known as flutes. Close to the lower margin, some glaciers accumulate so much debris beneath them that they actually glide on a bed of pressurized muddy till. As basal ice flows around a pronounced bedrock knob or a boulder lodged in the substrate, a cavity often forms in the ice on the lee side of the obstacle because of the high viscosity of the ice. Any pressurized muddy paste present under the glacier may then be injected into this cavity and deposited as an elongate tail of till, or flute. The size depends mainly on the size of the obstacle and on the availability of subglacial debris. Flutes vary in height from a few centimetres to tens of metres and in length from tens of centimetres to kilometres, even though very large flutes are generally limited to continental ice sheets.

Depositional Landforms of Continental Glaciers

Many of the deposits of continental ice sheets are very similar to those of valley glaciers. Terminal, end, and recessional moraines are formed by the same process as with valley glaciers, but they can be much larger. Morainic ridges may be laterally continuous for hundreds of kilometres, hundreds of metres high, and several kilometres wide. Since each moraine forms at a discreet position of the ice margin, plots of end moraines on a map of suitable scale allow the reconstruction of ice sheets at varying stages during their retreat.

In addition to linear accumulations of glacial debris, continental glaciers often deposit a more or less continuous, thin (less than 10 metres) sheet of till over large areas, which is called ground moraine. This type of moraine generally has a "hummocky" topography of low relief, with alternating small till mounds and depressions. Swamps or lakes typically occupy the low-lying areas. Flutes are a common feature found in areas covered by ground moraine.

Another depositional landform associated with continental glaciation is the drumlin, a streamlined, elongate mound of sediment. Such structures often occur in groups of tens or hundreds, which are called drumlin fields. The long axis of individual drumlins is usually aligned parallel to the direction of regional ice flow. In long profile, the stoss side of a drumlin is steeper than the lee side. Some drumlins consist entirely of till, while others have bedrock cores draped with till. The till in many drumlins has been shown to have a "fabric" in which the long axes of the individual rocks and sand grains are aligned parallel to the ice flow over the drumlin. Even though the details of the process are not fully understood, drumlins seem to form subglacially close to the edge of an ice sheet, often directly down-ice from large lake basins overridden by the ice during an advance. The difference between a rock drumlin and a drumlin is that the former is an erosional bedrock knob, whereas the latter is a depositional till feature.

Meltwater Deposits

Much of the debris in the glacial environment of both valley and continental glaciers is transported, reworked, and laid down by water. Whereas glaciofluvial deposits are formed by meltwater streams, glaciolacustrine sediments accumulate at the margins and bottoms of glacial lakes and ponds.

Glaciofluvial Deposits

The discharge of glacial streams is highly variable, depending on the season, time of day, and cloud cover. Maximum discharges occur during the afternoon on warm, sunny summer days, and minima on cold winter mornings. Beneath or within a glacier, the water flows in tunnels and is generally pressurized during periods of high discharge. In addition to debris washed in from unglaciated highlands adjacent to the glacier, a glacial stream can pick up large amounts of debris along its path at the base of the glacier. For this reason, meltwater streams issuing forth at the snout of a valley glacier or along the margin of an ice sheet are generally laden to transporting capacity with debris. Beyond the glacier margin, the water, which is no longer confined by the walls of the ice tunnel, spreads out and loses some of its velocity. Because of the decreased velocity, the stream must deposit some of its load. As a result, the original stream channel is choked with sediments, and the stream is forced to change its course around the obstacles, often breaking up into many winding and shifting channels separated by sand and gravel bars. The highly variable nature of the sediments laid down by such a braided stream reflects the unstable environment in which they form. Lenses of fine-grained, cross-bedded sands are often interbedded laterally and vertically with stringers of coarse, bouldery gravel. Since the amount of sediment laid down generally decreases with distance from the ice margin, the deposit is often wedge-shaped in cross section, ideally gently sloping off the end moraine formed at that ice position and thinning downstream. The outwash is then said to be "graded to" that particular moraine. In map view, the shape of the deposit depends on the surrounding topography. Where the valleys are deep enough not to be buried by the glaciofluvial sediments, as in most mountainous regions, the resulting elongate, planar deposits are termed valley trains. On the other hand, in low-relief areas the deposits of several ice-marginal streams may merge to form a wide outwash plain, or sandur.

If the ice margin stabilizes at a recessional position during glacial retreat, another valley train or sandur may be formed inside of the original one. Because of the downstream thinning of the outwash at any one point in the valley, the recessional deposit will be lower than and inset into the outer, slightly older outwash plain. Flat-topped remnants of the older plain may be left along the valley sides; these are called terraces. Ideally each recessional ice margin has a terrace graded to it, and these structures can be used in addition to moraines to reconstruct the positions of ice margins through time. In some cases where the glacier either never formed moraines or where the moraines were obliterated by the outwash or postglacial erosion, terraces are the only means of ice margin reconstruction.

Streams that flow over the terminus of a glacier often deposit stratified drift in their channels and in depressions on the ice surface. As the ice melts away, this ice-contact stratified drift slumps and partially collapses to form stagnant ice deposits. Isolated mounds of bedded sands and gravels deposited in this manner are called kames. Kame terraces form in a similar manner but between the lateral margin of a glacier and the valley wall. Glacial geologists sometimes employ the term kame moraine to describe deposits of stratified drift laid down at an ice margin in the arcuate shape of a moraine. Some researchers, however, object to the use of the term moraine in this context because the deposit is not composed of till.

In some cases, streams deposit stratified drift in subglacial or englacial tunnels. As the ice melts away, these sinuous channel deposits may be left as long linear gravel ridges called eskers. Some eskers deposited by the great ice sheets of the Pleistocene can be traced for hundreds of kilometres, even though most esker segments are only a few hundred metres to kilometres long and a few to tens of metres high.

Kettles, potholes, or ice pits are steep-sided depressions typical of many glacial and glaciofluvial deposits. Kettles form when till or outwash is deposited around ice blocks that have become separated from the active glacier by ablation. Such "stagnant" ice blocks may persist insulated under a mantle of debris for hundreds of years. When they finally melt, depressions remain in their place, bordered by slumped masses of the surrounding glacial deposits. Many of the lakes in areas of glacial deposition are water-filled kettles and so are called kettle lakes. If a sandur or valley train contains many kettles, it is referred to as a pitted outwash plain.

Glaciolacustrine Deposits

Glacial and proglacial lakes are found in a variety of environments and in considerable numbers. Erosional lake basins have already been mentioned, but many lakes are formed as streams are dammed by the ice itself, by glacial deposits, or by a combination of these factors. Any lake that remains at a stable level for an extended period of time (e.g., hundreds or thousands of years) tends to form a perfectly horizontal, flat, terracelike feature along its beach. Such a bench may be formed by wave erosion of the bedrock or glacial sediments that form the margin of the lake, and it is called a wave-cut bench. On the other hand, it may be formed by deposition of sand and gravel from long-shore currents along the margin of the lake, in which case it is referred to as a beach ridge. The width of these shorelines varies from a few metres to several hundred metres. As the lake level is lowered due to the opening of another outlet or down cutting of the spillway, new, lower shorelines may be formed. Most former or existing glacial lakes (e.g., the Great Salt Lake and the Great Lakes in North America) have several such shorelines that can be used both to determine the former size and depth of now-extinct or shrunken lakes and to determine the amount of differential postglacial uplift because they are now tilted slightly from their original horizontal position.

Where a stream enters a standing body of water, it is forced to deposit its bedload. The coarser gravel and sand are laid down directly at the mouth of the stream as successive, steeply inclined foreset beds. The finer, suspended silt and clay can drift a bit farther into the lake, where they are deposited as almost flat-lying bottomset beds. As the sediment builds out farther into the lake (or ocean), the river deposits a thin veneer of subhorizontal gravelly topset beds over the foreset units. Because the foreset–topset complex often has the shape of a triangle with the mouth of the stream at one apex, such a body of sediment is called a delta. Many gravel and sand pits are located in deltas of former glacial lakes.

Chile: glacial lake

The flat-lying, fine-grained bottomset beds of many large former glacial lakes filled in and buried all of the pre-existing relief and are now exposed, forming perfectly flat lake plains. Cuts into these sediments often reveal rhythmically interbedded silts and clays. Some of these so-called rhythmites have been shown to be the result of seasonal changes in the proglacial environment. During the warmer summer months, the meltwater streams carry silt and clay into the lakes, and the silt settles out of suspension more rapidly than the clay. A thicker, silty summer layer is thus deposited. During the winter, as the surface of the lake freezes and the meltwater discharge into it ceases, the clays contained in the lake water slowly settle out of suspension to form a thin winter clay layer. Such lacustrine deposits with annual silt and clay "couplets" are known as varves.

Periglacial Landforms

In the cold, or periglacial (near-glacial), areas adjacent to and beyond the limit of glaciers, a zone of intense freeze-thaw activity produces periglacial features and landforms. This happens because of the unique behaviour of water as it changes from the liquid to the solid state. As water freezes, its volume increases about 9 percent. This is often combined with the process of differential ice growth, which traps air, resulting in an even greater increase in volume. If confined in a crack or pore space, such ice and air mixtures can exert pressures of about 200,000 kilopascals (29,000 pounds per square inch). This is enough to break the enclosing rock. Thus freezing water can be a powerful agent of physical weathering. If multiple freeze-and-thaw cycles occur, the growth of ice crystals fractures and moves material by means of frost shattering and frost heaving, respectively. In addition, in permafrost regions where the ground remains frozen all year, characteristic landforms are formed by perennial ice.

Felsenmeers, Talus and Rock Glaciers

In nature, the tensional strength of most rocks is exceeded by the pressure of water crystallizing in cracks. Thus, repeated freezing and thawing not only forms potholes in poorly constructed roads but also is capable of reducing exposed bedrock outcrops to rubble. Many high peaks are covered with frost-shattered angular rock fragments. A larger area blanketed with such debris is called a felsenmeer, from the German for "sea of rocks." The rock fragments can be transported downslope by flowing water or frost-induced surface creep, or they may fall off the cliff from which they were wedged by the ice. Accumulations of this angular debris at the base of steep slopes are known as talus. Owing to the steepness of the valley sides of many glacial troughs, talus is commonly found in formerly glaciated mountain regions. Talus cones are formed when the debris coming from above is channelized on its way to the base of the cliff in rock chutes. As the talus cones of neighbouring chutes grow over time, they may coalesce to form a composite talus apron.

In higher mountain regions, the interior of thick accumulations of talus may remain at temperatures below freezing all year. Rain or meltwater percolating into the interstices between the rocks freezes over time, filling the entire pore space. In some cases, enough ice forms to enable the entire mass of rock and ice to move downhill like a glacier. The resulting massive, lobate, mobile feature is called a rock glacier. Some rock glaciers have been shown to contain pure ice under a thick layer of talus with some interstitial ice. These features may be the final retreat stages of valley glaciers buried under talus.

Permafrost, Patterned Ground, Solifluction Deposits and Pingos

Permafrost is ground that remains perennially frozen. It covers about 20–25 percent of the Earth's land surface today. The "active layer" of soil close to the surface of permafrost regions undergoes many seasonal and daily freeze-thaw cycles. The constant change in the volume of water tends to move the coarser particles in the soil to the surface. Further frost heaving arranges the stones and rocks according to their sizes to produce patterned ground. Circular arrangements of the larger rocks are termed stone rings. When neighbouring stone rings coalesce, they form polygonal stone nets. On steeper slopes, stone rings and stone nets are often stretched into stone stripes by slow downhill motion of the soggy active layer of the permafrost. In other areas, patterned ground is formed by vertical or subvertical polygonal cracks, which are initiated in the soil by contraction during extremely cold winters. During the spring thaw of the active layer, water flows into these cracks, freezes, and expands. This process is repeated year after year, and the ice-filled cracks increase in size. The resulting ice wedges are often several metres deep and a few tens of centimetres wide at the top. Along the sides of ice wedges, the soil is deformed and compressed. Because of this disturbance and sediment that may be washed into the crack as the ice melts, relict patterned ground may be preserved during a period of warmer climate long after the permafrost has thawed. Today, relict patterned ground that formed during the last ice age exists more than 1,000 kilometres to the south of the present limit of permafrost.

When the active layer of permafrost moves under the influence of gravity, the process is termed gelifluction. The soft flowing layer is often folded and draped on hillsides and at the base of slopes as solifluction, or gelifluction, lobes.

In some permafrost areas, a locally abundant groundwater supply present at a relatively shallow depth may cause the exceptional growth of ice within a confined area. The sustained supply of liquid water results in the expansion of an increasingly large, lens-shaped ice body. These conical mounds, or pingos, may be several tens of metres high and hundreds of metres in diameter.

Coastal Processes and Landforms

Profile of Coastal Zones

When we talk about the coastal zone, the first thing to know about is the coastline.

A Coastline represents the boundary between the continental land masses and the oceanic water masses. A Coastal zone is the transition zone between terrestrial habitat and the marine habitat. It is the interface between land and oceanic water. Coastal belts may be very wide or narrow. They also vary with reference to their slope, beach profile, rock types, climate and vegetation. The climate of a coast is basically controlled by the land and sea breezes. The climate is also controlled by the humidity of generated by the marine waters.

If we look at the profile of a sea coast, we can see a shoreline belonging to the landward side. The width of this shore may vary from place to place. Along the coast, there are two distinct zones observable in a day. They are the high tide line and the low tide line. The average water level between the high tide and the low tide is the mean sea level. When there is a severe storm, the water line may come well above the high tide line in a beach.

Coastal belts are further divided into three divisions as,

1. Backshore region

2. Foreshore region

3. Offshore region

The backshore is inland of the inter-tidal zone and is usually above the influence of the waves. The nearshore (sometimes called the breaker zone) is where the waves break; the offshore zone is further out to sea and is beyond the influence of the waves.

1. Backshore

The Backshore region represents the beach zone starting from the limit of storm wave, above high tide shoreline. This zone includes a wave cut terrace and a storm scarp. Beach is the basic area where much of the geological processes happening. Beach is the sloping portion of the coast normally existing below the berms. This area is partly exposed by the backwash of waves (swash zone).

Swash zone is the place where the waves backwash the materials. It is the place where up and down movement of beach materials take place.

2. Foreshore

The Foreshore region is the region between high tide water zone and low tide water zone. It includes a Beach face and a beach terrace. The surf zone exists above beach terrace. At the end of the surf zone, the breaker zone starts. The foreshore may be a sandy foreshore, shingle foreshore, muddy foreshore or a rocky foreshore.

The Breaker zone is the area where the incoming waves become unstable, raising to a peak and breaking down.

Breaker zone is an important zone within which waves approaching the coastline commence breaking. The breaker zone is also part of the surf zone. The Surf zone is an important zone where the waves of translation occur after the waves break. Sand Bars are created, inside the waters, along the zone of wave breakers. The moving water masses shape the excess quantities of detritus sediments into sorted and layered deposits.

Long shore currents occur in this zone, which run parallel to the coastline.

3. Offshore

The Offshore region represents the zone of oceanic shallow water zone extending fully inside the continental shelf. It begins after breaker zone. At the base, it includes the longshore troughs and longshore bars.

Waves, Tides and Currents

When we go to a coastal region, we hear the continuous sounds of the waves and their movements on the coastal belts.

Waves are the powerful forces acting on the coastal zones. They are very dynamic systems with reference to space and time. Waves are capable of doing both constructive and destructive processes.

Waves get their energy from the wind. As the wind blows over the surface of the sea, it creates friction. This frictional drag causes water particles to rotate and their energy is transferred forward in the form of a wave. Whilst the water moves forward, the water particles return to their original position.

As a wave reaches the shallow water region, friction between the sea bed and the bottom of the wave causes it to slow down. Its shape becomes more elliptical rather than circular. The top of the wave, which is called as the crest, is not affected by the friction, and it becomes steeper until it eventually breaks.

When the wave breaks, the water rushes up towards the elevated parts of the beach. This movement is called as the swash. Then it comes down slowly backwards. The movement of the same water, back down the beach, is called as the backwash.

Waves have a strength to act. There are three main factors that affect the strength of a wave:

1. The strength and speed of the wind.

2. The duration of the wind - this is the length of time for which the wind has blown.

3. The fetch - this is the distance over which the wind has blown.

The rise and fall of oscillatory waves in an open water reflects the circular motion of water particles. Swells are smooth, rounded waves that travel outward from a storm center. Sea waves are classified into two types on the basis of the depth of oceanic waters. They are:

a) Oscillatory waves (these are the waves in deep water) and

b) Translatory waves (these are the waves in shallow water).

Waves are one of the most significant forces in shaping the coastline.

Based on their actions, two main types of waves have been recognised.

They are:

a) Constructive waves and

b) Destructuve waves.

Constructive waves are low energy waves that tend to arrive at the coast at a rate of less than 8 waves per minute. Constructive waves are small in height. They have a strong swash and a weak backwash. This means that constructive waves tend to deposit material and build up a beach.

Destructive waves have much higher energy and tend to arrive at the coast at a rate of more than 8 per minute. They are much larger in height than constructive waves, often having been caused by strong winds and a large fetch. Destructive waves have a weak swash but a strong backwash so they erode the beach by pulling sand and shingle down the beach as water returns to the sea. This means that less beach is left to absorb wave energy.

Seismic sea Waves called Tsunami are yet another force of oceans. A tsunami originates from the deep oceans and reaches the continents in the form of massive strong waves. These are devastating water wave generated by an undersea earthquake.

Coastal Geomorphic Processes

Oceans are bodies of dynamic water masses. Sea waves are powerful geological agents, acting from the shorelines to the coastal belts.

Vertical and horizontal movements of water continue to happen both at the surface and at depth at all times. Over a period of time, wave action in the surf zone will tend to plane off the entire zone. This process is known as marine planation. This is a slow process.

There are so many other features formed along the coastline due to various hydrodynamic actions of waves on the sea side and aerodynamic actions wind on the landside.

Sea waves can erode, transport and deposit the marine sediments based on various factors and processes. Erosion, transportation and deposition happen on both sides of the shoreline. Coastal rocks like cliffs are also subjected to wave actions. Sea cliffs are very unique features seen in some places.

Processes of Coastal Erosion

The combined effect of waves, currents and tides result in a variety of gradational processes acting in the coastal zone. Coastal erosion happens in the form of:

a) Hydraulic action,

b) Corrosion (or) abrasion,

c) Attrition,

d) Corrosion (or) solution, and

   e) Water pressure.

Hydraulic action is the impact of moving water on the coastal rocks. It is caused by the direct impact of waves on the coasts. Enormous pressures can build as water and air are compressed into the rock fractures.

The most important one is abrasion. Abrasion (or) corrosion is a kind of erosion happening with the help of tools of erosion. In water suspension coarse sands, pebbles, cobbles and boulders are used by the waves to attack the coastal rocks.

Attrition is a process in which mechanical tear and wear can break any rock mass into fragments. Mutual collision effected by backwash and rip currents are powerful tools of coastal erosion.

Corrosion (or) Solution is the chemical alteration of rocks which are soluble and due to their contact with the seawater. Solution is locally important especially where soluble rock is exposed along the shore.

Due to periodic wetting and drying a wide range of chemical processes happen on the coastal rocks which lead to both physical disintegration and chemical decomposition.

Alternate freeze and thaw can also make these rocks to be easily attacked by the waves. A good number of coastal features are formed by the action of these sea waves.

Coastal sediments are subject to multiple episodes of erosion, transportation and deposition, though a net seaward transport takes place on a global scale. The deep ocean floor becomes the resting place for terrestrial sediment eroded from the land.

Beach drifting transports sand grains along the beach as waves strike the shore at an oblique angle. Sediment is carried landward when water rushes across the beach as swash. Sediment is carried back toward the ocean as backwash. The continual up rush and backwash carries sand in a zig-zag like movement along the shore.

Erosional Landforms

Landforms of coastal regions are classified into two major groups as erosional landforms and depositional landforms.

The notable erosional landforms of the coasts are:

   a) Sea cliffs

   b) Sea caves

   c) Sea Arches

   d) Sea stacks

   e) Wave-cut notches

   f) Wave-built terraces

The most widespread landforms of erosional coasts are sea cliffs. Wave erosion undercuts steep shorelines creating coastal cliffs. A sea cliff is a vertical precipice created by waves crashing directly on a steeply inclined slope. These very steep to vertical bedrock cliffs range from only a few metres high to hundreds of metres above sea level.

Their vertical nature is the result of wave-induced erosion near sea level and the subsequent collapse of rocks at higher elevation.

Hydraulic action, abrasion, and chemical solution all work to cut a notch at the high water level near the base of the cliff.

Constant undercutting and erosion causes the cliffs to retreat landward. Sea caves form along lines of weakness in cohesive but well-jointed bedrock. Sea caves are prominent headlands where wave refraction attacks the shore.

A sea arch forms when sea caves merge from opposite sides of a headland. If the arch collapses, a pillar of rock remains behind as a sea stack.

Seaward of the retreating cliffs, wave erosion forms a broad erosional platform called a wave-cut bench or wave-cut platform. After the constant grinding and battering, eroded material is transported to adjacent bays to become beaches or seaward coming to rest as a wave-built terrace.

Depositional Landforms

Eroded sediments along the coasts are transported as drifts. Longshore Drift is one important mechanism. Longshore Drift are powerful geomorphic agents. They can erode, transport and deposit coastal sediments.

Longshore drift erodes and deposits sand masses continuously along the beach. The sand that is removed from one point along the beach is replaced by sand eroded from some other zones.

Longshore drift consists of the transportation of sediments like clay, silt, sand and shingle. The drift happens along a coast at an angle to the shoreline. This is mainly dependent on the prevailing wind direction, swash and backwash.

This process occurs in the littoral zone, and in or close to the surf zone. The process is also known as longshore transport or littoral drift.

Littoral transport is the term used for the transport of non-cohesive sediments, i.e. mainly sand, along the foreshore and the shoreface due to the action of the breaking waves and the longshore current.

Tides

Tides are Routine Coastal Processes

Nearly all marine coastlines experience the rhythmic rise and fall of sea level called tides. The daily oscillation in ocean level is a product of the gravitational attraction of the Moon and Sun on Earth's oceans and it varies in degree worldwide. Tidal action is an important force behind coastal erosion and deposition as the shoreline migrates landward and seaward.

Tidal Currents are responsible for mechanical sorting of sediments under the water. During a high tide water moves landward as a flood current. During low tide water recedes seaward as an ebb current.

The notable depositional coastal landforms are:

a) Beaches

b) Spits and bars

c) Tombolo

d) Barrier islands

e) Mud Flats

Beaches

A beach is an area of sediment accumulation. They are exposed to wave action along the coast. Beaches morphology changes from season to season.

There are two basic beach types:

One is called as dissipative beach and the other one is called as reflective beach. Together with the intermediate types, there are six major microtidal beach types. The reflective beach occurs when conditions are calm and the sediment is coarse. There is no surf zone. The waves flow upon the reflective beaches. It reflects a major part of the incoming wave.

When bigger waves cut back a beach and spread out its sediments to form a surf zone, the reflective beaches create a series of intermediate types.

When wave action is very strong and sediment particle size is fine, the dissipative beach type is created.

This type has a flat and maximally eroded beach.

The sediments are stored in a broad surf zone that may have multiple sandbanks parallel to the beach.

The intermediate types are characterized by high temporal variability, sand storage both on the beach and in the surf zone and sandbanks and troughs.

Beaches are classified into three categories as high, low and moderate energy beaches. Normally, high energy conditions prevail during summer months. The wave heights are normally expected to increase after the onset of monsoons. These produce significant changes in the beach morphology.

Spits and Bars

A sand spit is one of the most common coastal landforms. A sand spit is a linear accumulation of sediment that is attached to land at one end. Sand carried parallel to shore by longshore drift may eventually extend across a bay or between headlands especially where water is relatively calm.

Spits are typically elongated, narrow features built to several meters high above sea level by the action of wind and waves.

Spits often form when wave energy decreases as a result of wave refraction in a bay. When the wave energy is dissipated, it will cause the sediment to accumulate, due to the loss of ability to transport the sediments by water.

The term bar refers to a long narrow sand embankment formed by wave action. Littoral drift from an island may form a tombolo, which is a sand bar connecting the island with the mainland.

Spits can extend across the mouth of a bay, but wave action is usually strong enough to wash sand out to sea or be deposited in the embayment. They may curve into the bay or stretch across connecting to the other side as a baymouth bar. When the bay is closed off by a bar it becomes a lagoon.

Simple spits consist of narrow finger of sand with a single dune ridge that elongates in the downdrift direction.

Double spits can form if drift transports sand in two directions across and inlet, or if a bay mouth barrier is cut by a tidal channel. Wave refraction at the end of a spit will transport sand to form a recurved spit. Complex spits form when a plentiful supply of sediment is transported by both the ocean and bay currents. Multiple lines of dunes can be formed by wind transport of sand across the spit.

Tombolo

A tombolo is a depositional landform in which an island is attached to the mainland by a narrow piece of land such as a spit or bar. Tombolos are formed by wave refraction.

Coastlines paralleled by offshore narrow strips of sand dunes, salt marshes and beaches are known as barrier islands. A variety of barrier-related features could be seen along the shoreline.

Barrier Spits and Islands

Barriers that connect headlands together along the outer reaches of an embayment are called baymouth barriers.

Barrier spits are beaches that are attached at one end to their source of sediment.

Capes are barrier islands that project into the open sea to form a right angle shoreline. These are generally large features that are exposed to wave attack on each side, but one side is accreting while the other is eroding. This produces a distinctive series of truncated dune ridges.

Mud Flats

Mud flats are formed due to the action of tidal currents. They contain silt and clays. They are exposed during low tides and are covered during high tides. In some of the exposed mud flats, after a full retreat of a sea level, plants grow in these mudflats forming salt marshes.

200   Introductory Geography

In addition to these some other features are also located in the coastal areas:

Estuaries

An estuary is a coastal wetland where freshwater from runoff of a river and saltwater from tides of the seas and oceans mix together. Most of the large rivers in the world do not empty their waters abruptly into the seas. They merge with the sea in a transitional basin-like area near their mouths called as an estuary.

Deltas

These are bodies of sediments deposited by the rivers when they confluence with the seas. Deltas build outward from the shoreline at river mouths.

There are three kinds of deltas as:

a) Wave-dominated Deltas

b) Tide-dominated Deltas

c) River-dominated Deltas

References

- Landform: niu.edu, Retrieved 9 July, 2019

- Erosion-deposition-running-water-ground-water: clearias.com, Retrieved 19 April, 2019

- Glacial-landform, science: britannica.com, Retrieved 16 August, 2019

- Coastal-Processes-and-Landforms: researchgate.net, Retrieved 9 May, 2019

Chapter 6

Population and Migration

The total number of people which reside in a particular place over a period of time constitute the population of that place. The study of populations involves studying its composition, density and growth. The chapter closely examines these key concepts of population and migration to provide an extensive understanding of the subject.

Population and its Composition

Population – It is number of people living in a region between two point of time.

"Population is the total process of collecting, compiling, analysing or otherwise disseminating demographic, economic and social data pertaining, at a specific time, to all persons in a country or a well-defined part of a country" Population definition census of India.

Population composition - is the description of a population according to characteristics such as age and sex.

Sex composition (Sex ratio) - It is the ration between the number of women and men in the population.

The sex ratio gives important information about the status women in the country The regions where Gender discrimination is rampant sex ratio is unfavorable for women. In such areas female foeticide, female in fanticide and domestic violence prevalent.

Different countries use different method to calculate sex ratio .In some countries it is calculated by using the formula (male population/female population)*1000

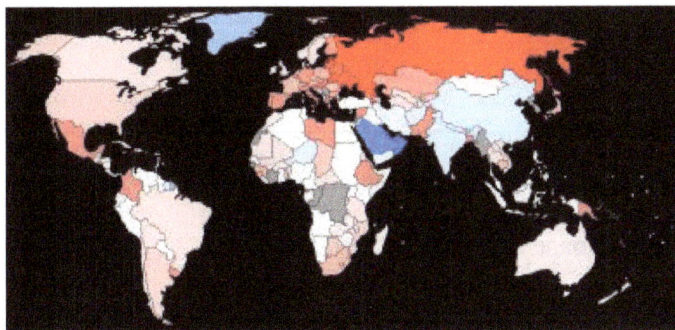

Blue shade show less favorable countries.

The country have a high male mortality rate than female. Lowest sex ratio is in Qatar where it is about 311male 100 female. This imbalance of sex ratio due to the high migration of male skilled workers to the Qatar.

There are over 139 nation's has favorable sex ration and 79 countries have less favorable condition.

The deficit of male population in European countries is due to out migration of male.

Age Structure

Age distribution of human populations: A good way to determine how many people at what age groups live in an area is to use an age pyramid, which measures age structure.

It is one of important indicator of population composition working population – It is the number of people between the age group of 15- 59 Dependent population – People below 14 years and 60 and above are included in this classification. Dependent population requires more expenditure in health care and other amenities.

Age sex pyramid – this graph show number of male and female in different age group In the left side of pyramid graph shows percentage of male and right side percentage shows of female

Shape of pyramid shows the characteristics of the population.

Expanding Population

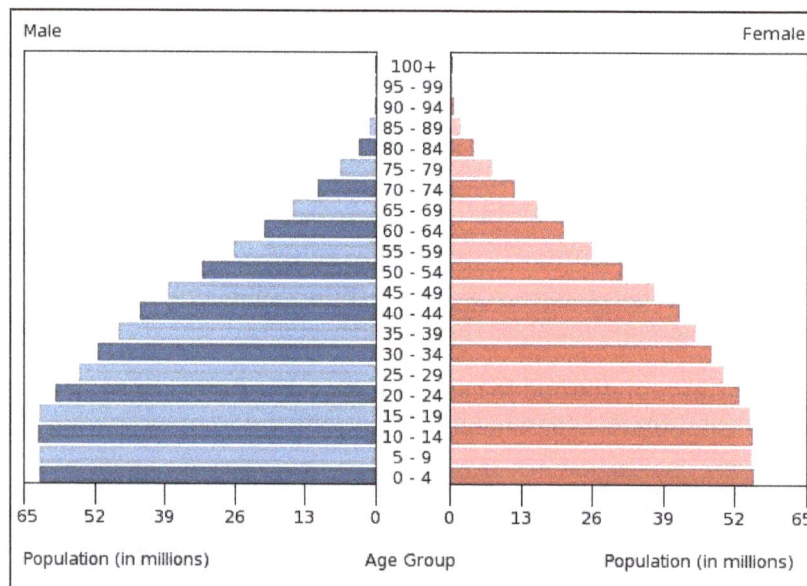

1.   This is triangular shaped pyramid graph.

2.   This graph have wide base tapering toward the top.

3.   This graph shows the population type of less developed countries.

4.   This graph shows high birth rate.

5.   This graph shows that the dependent population is more than that of working populations.

6.   Nigeria, Bangladesh.

Constant Population

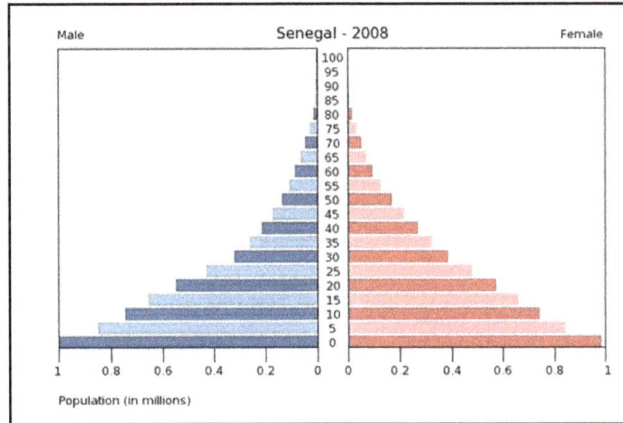

1. This graph represent age structure pyramid of developed contrary.

2. The graph resembles a bell structure and tapered towards top.

3. Birth rate and death rate are almost equal leading to near constant population.

4. Working population is more than that of dependent population.

5. Near-stationary population pyramids display somewhat equal numbers or percentages for almost all age groups.

6. Smaller figures are still to be expected at the oldest age groups.

7. The age-sex distributions of some European countries, especially Scandinavian ones, will tend to fall into this category.

Declining Population

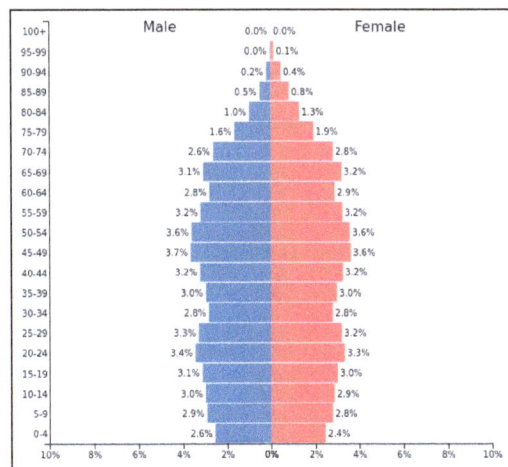

1. This graph shows the population structure of Japan.

2. The Graph have narrow base and tapered toward the top.

3. This graph is shows low birth rate and death rate.

4. the population growth is zero or - ve growth.

5. Population decrease is due to high marriage age.

Population Density

Population density is a term that expresses how crowded (or uncrowded) a particular piece of land is. Governments keep track of population density so that the people in charge have an idea of how many citizens live in a particular area and can provide services accordingly. In fact, the U.S. Census Bureau keeps close track of the population density in many areas of the United States so that it can coordinate with the National Hurricane Center to explain how many people and businesses a storm might affect.

Find the Population for your Selected Area

Make a fraction with the correct units for population density. Scientists and sociologists usually express population density as a number of people per unit of area. Therefore, to create a correct fractional expression, you would need to have the number of people in the numerator (on the top of the fraction) and the number of square miles/meters/acres in the denominator (on the bottom).

For example, a population of 8,341 people living in an area of 25 square miles would be expressed as a fraction: 8341 / 25.

Reduce the fraction to a unit area so that the final result is the number of people in one acre or square mile or square meter, for example. Another way to think of this is to carry out the division represented by the fraction. If you wanted to calculate the population density of a town that is 25 square miles and has a population of 8,341 people, you would divide 8,341 by 25 to get 333.64 people per square mile.

Round your answer from Step 5 up or down to a whole number, if necessary. The population density is simply an average, so you can end up with a decimal; however, the real population is composed of whole people, so rounding will express your answer in whole people as well.

Factors Influencing Population Distribution

It is, however, not to suggest that population distribution on the earth surface is determined by physical factors alone, for within the broad framework of physical attractions and constraints, cultural factors strongly influence the way mankind is distributed over the earth. Thus, apart from physical factors, numerous social, demographic, economic, political and historical factors affect population distribution.

These factors operate not in isolation but in combination with each other. One cannot, therefore, isolate the influence of any one factor on population distribution. Further, the interplay between these determinants is generally very complex. The primary task of a population geographer, therefore, is to explain the irregularities in population distribution in terms of the influences of all these factors as an integral part of a dynamic process.

Physical Factors

Physical factors that affect population distribution include altitude and latitude, relief, climate, soils, vegetation, water and location of mineral and energy resources. It is important to note that most of the physical factors influence population distribution only indirectly through climatic conditions.

The influences of latitude and altitude on population distribution cannot be separated from one another. High altitude in general imposes an ultimate physiological limit upon human existence due to reduced atmospheric pressure and low oxygen content. Therefore, very few permanent settlements can be seen in the lofty mountains of the world at a height above 5,000 metres. Staszewski, in his exhaustive analysis of the vertical distribution of population, has shown that both numbers and densities in different parts of the world decline with increasing altitude.

According to him, a little more than 56 per cent of the world's population lives within 200 metres from the sea level, and over 80 per cent within 500 metres. However, in low latitude areas, which are otherwise hot and less favorable, high altitude provides suitable conditions for human habitation. Mountains in Africa and Latin America are much healthier than plains, and large cities have sprung up at high altitude. La Paz, the highest city in the world (3,640 m) and the capital of Bolivia, owes its existence to this factor. As against this, in the high latitude areas, it becomes extremely difficult to live beyond a few hundred metres from sea level. It is in this context that a famous population geographer has referred to "mountains that attract and mountains that repel".

Relief features also play an important role in influencing population distribution. The influence of altitude has already been noted. Among the other aspects of relief features which affect human habitation are general topography, slope and aspect. The main concentrations of human population are confined to the areas marked with flat topography. Rugged and undulating topography restricts the condensation of human population in any area.

Abrupt changes in the density of population can be seen on the world map of population distribution where plains meet mountain ranges. Rising Himalayas, thus, mark the northern limit of dense population in the Ganga plain. Similarly, the Deccan plateaus with rugged and undulating topography appear distinct from the plains in respect of population concentration. In the mountainous areas valleys provide suitable locations for human settlements. Likewise, sun-facing slopes provide favorable locations for the emergence and growth of settlements.

This is particularly true in the temperate and other high latitude areas where insolation is very important. The river valleys may promote or restrict human settlements depending upon other geographic conditions. In Egypt, nearly 98 per cent of the population is concentrated forming a ribbon along the Nile River. As against this, in tropical swamps and dissected plateaus, river valleys tend to repel population.

Of all the geographic influences on population distribution, climatic conditions are perhaps the most important. Climate affects population distribution both directly as well as indirectly through its effects on soil, vegetation and agriculture that have direct bearings on the pattern of population distribution. Moreover, other physical factors like latitude and altitude also operate on population distribution through climatic conditions.

Although climatic optima are difficult to define, extremes of temperature, rainfall and humidity certainly limit the concentration of population in any part of the earth. In the Northern Hemisphere, extreme cold conditions in the high latitude areas have prevented human habitation. Likewise, extremely high temperature and aridity in the hot deserts of the world restrict human habitability. Some of the geographers in past have, therefore, gone to the extent of claiming a deterministic relationship between climate and population distribution.

It should, however, be noted that man has ability to adapt himself to different climatic conditions. This explains a high density in the tropics, which are otherwise marked with extremes of climatic conditions. Progress in science and technology has greatly augmented man's ability to adapt to different climatic conditions. Though limited in magnitude, the peopling of the Alaska and Siberia during the last century owes to the scientific and technological advancements.

The cases of Java and the Amazon basin also serve to refute deterministic stance of relationship between climate and population distribution. Though, both of them experience equatorial type of climate, they differ markedly from one another in terms of population density. While Java is one of the most densely parts of the world, the Amazon basin is marked with a very sparse population.

Similarly, the quality of soils exerts an undeniable influence on the distribution of world population. The fertile alluvial and deltaic soils can support dense populations. Thus, most of the major concentrations of populations in the world are located in the river valleys and deltas. Great civilizations of the world have almost invariably flourished on good fertile alluvial soils. Similarly, the chernozems of steppe grasslands and rich volcanic soils can support dense population.

On the other hand, the leached soils of temperate lands, the podsols, which are very poor in terms of fertility, can support only a sparse population. In Canada, for instance, marked difference can be noticed in population concentration between areas of clayey soils and podsol soils.

It is important to note that the influence of soils cannot be viewed in isolation, that is, soils influence population distribution in association with other physical factors, mainly climate. Moreover, progress in technology can alter the effectiveness of soil types on population concentration to a greater extent. Application of modern technologies during the recent times has greatly enhanced the profitability of cultivation in many areas of the world, which were hitherto not suitable for cultivation.

Such areas have, thus, attracted population during the recent past. In association with climatic conditions, varying soil types give rise to variety of vegetation cover on the earth surface. These, in turn, provide contrasting environment for a variety of agricultural activities, and hence, lead to different population density. Tropical forests, savanna, tundra and taiga provide different media for human occupation and concentration.

Location of mineral and energy resources has led to dense population concentration in many parts of the world, which otherwise do not provide suitable conditions for human habitation. Large towns have grown up in inaccessible and extremely inhospitable areas such as deserts, Polar Regions or in the midst of forests where precious minerals and metals have been found.

Kalgoorlie, a gold mining town in the Australian deserts, is a very good example in this regard. Likewise, several other examples can be cited from elsewhere in the world including Canada, the USA and Russia. Location of coal, the most important fuel in the nineteenth and early twentieth centuries was the main factor behind industrial conurbation and dense population concentration in Western Europe.

However, the influence of mineral and energy resources on population distribution depends upon a wide range of social and economic factors such as market demand, capital for development, availability of labour supply and transportation network. It is, therefore, important to note that the influences of all the physical factors outlined above operate through a series of economic, social and political factors in the area concerned.

Economic, Political and Historical Factors

Population distribution and density in an area depends to a large extent on the type and scale of economic activities. Same geographic conditions provide different opportunities for people with different types and scale of economic activities. Technological and economic advancement can bring about significant changes in population distribution of an area. For instance, the Prairies of North America offered different opportunities for the Indians with their hunting economy, the nineteenth century ranchers, the later settled agriculturist and finally the modern industrialized and largely urbanized society.

Each stage in economic development was marked with profound changes in population density and distribution in the region. Industrialization and discovery of new sources of minerals and energy resources have, throughout human history, brought about redistribution of population through migration. In the pre-industrial agricultural societies, population distribution often fairly evenly distributed responds to the nature of crops grown and their relationship to physical conditions.

The industrial revolution has resulted in considerable change in population distribution in many parts of the world. Dense population concentration has replaced long established pattern of dispersal and generally even distribution. Initially, sources of energy and mineral resources became the force of industrial growth and population concentration. Improved transport network, growing spatial mobility of labour and increasing trade in the wake of economic and technological advancements have led to decline in the importance of place bound industries.

Growing commercial activities, for instance, in the developing world, accompanied by improvements in transport network, have resulted in considerable redistribution of population and emergence of mega urban centres. It is aptly said that increasing complexity and diversification of economic activities, the world over, have led to a new pattern of population distribution.

During the more recent times, government policies and political factors have emerged as an important determinant of population patterns. With increasing state control over economic activities, government policies have led to a significant change in the patterns of population distribution in several parts of the world. In the erstwhile USSR, facilitated by advances in science and technology, population was directed to parts of Siberian plains, which were hitherto not suitable for human habitation. Likewise, in China, planned colonization of the interior, encouraged by the communist government, has resulted in significant change in population patterns.

In the late 1960s and 1970s, some 10 to 15 million people in the country were forcibly relocated to the rural communes in order to ease pressure on urban employment. Examples of government inducements encouraging migration to new areas can be cited from several developed countries of the West as well. In addition to government policies, political events have also caused redistribution of population throughout human history.

Wars have forced a great number of people to migrate from one region to another all over the world. Post-partition redistribution between India and Pakistan, or displacement of several million Sudanese as a result of civil war, or expulsion of Asians from Uganda in the early 1970s are some of the instances of how political events can cause changes in population patterns.

Apart from the factors discussed above, historical processes should also be taken into account while explaining the patterns of population distribution. Duration of human settlements is an important determinant of the magnitude of population concentration in any area. Most of the densely populated areas of the world have a very long history of human habitation, while sparse population in certain areas can in part be explained in terms of its recent habitation.

It should, however, not be concluded that the highest densities are always to be found in areas with the longest history of habitation. There are several instances of formerly prosperous and densely populated areas, which are now only sparsely populated. Parts of North Africa and Mesopotamia, the Yucatan peninsula and eastern Sri Lanka are some such examples. Based on this, some scholars have talked about cycle of occupation, whereby size and densities of population first increase and then decline only to be followed by another cycle of increase.

Population Growth

Population growth refers to change in the size of a population—which can be either positive or negative—over time, depending on the balance of births and deaths. If there are many deaths, the world's population will grow very slowly or can even decline. Population growth is measured in both absolute and relative terms. Absolute growth is the difference in numbers between a population over time; for example, in 1950 the world's population was 4 billion, and in 2000 it was 6 billion, a growth of 2 billion. Relative growth is usually expressed as a rate or a percentage; for example, in 2000 the rate of global population growth was 1.4 percent (or 14 per 1,000). For every 1,000 people in the world, 14 more are being added per year.

For the world as a whole, population grows to the extent that the number or rate of births exceeds the number or rate of deaths. The difference between these numbers (or rates) is termed "natural increase" (or "natural decrease" if deaths exceed births). For example, in 2000 there were 22 births per 1,000 population (the number of births per 1,000 population is termed the "crude birth rate") and 9 deaths per 1,000 population (the "crude death rate"). This difference accounts for the 2000 population growth rate of 14 per 1,000, which is also the rate of natural increase. In absolute numbers, this means that approximately 78 million people—or about the population of the Philippines—is added to the world each year. For countries, regions, states, and so on, population growth results from a combination of natural increase and migration flows. The rate of natural increase is equivalent to the rate of population growth only for the world as a whole and for any smaller geographical units that experience no migration.

Populations can grow at an exponential rate, just as compound interest accumulates in a bank account. One way to assess the growth potential of a population is to calculate its doubling time—the number of years it will take for a population to double in size, assuming the current rate of population growth remains unchanged. This is done by applying the "rule of seventy"; that is, seventy divided by the current population growth rate (in percent per year). The 1.4 percent global population growth rate in 2000 translates into a doubling time (if the growth rate remains constant) of fifty-one years.

Components of Population Change

The crude birth rate (CBR) and crude death rate (CBR) are statistical values that can be used to measure the growth or decline of a population.

The crude birth rate and crude death rate are both measured by the rate of births or deaths respectively among a population of 1,000. The CBR and CDR are determined by taking the total number of births or deaths in a population and dividing both values by a number to obtain the rate per 1,000.

For example, if a country has a population of 1 million, and 15,000 babies were born last year in that country, we divide both the 15,000 and 1,000,000 by 1,000 to obtain the rate per 1,000. Thus the crude birth rate is 15 per 1,000.

The crude birth rate is called "crude" because it does not take into account age or sex differences among the population. In our hypothetical country, the rate is 15 births for every 1,000 people, but the likelihood is that around 500 of those 1,000 people are men, and of the 500 who are women, only a certain percentage are capable of giving birth in a given year.

Birth Trends

Crude birth rates of more than 30 per 1,000 are considered high, and rates of less than 18 per 1,000 are considered low. The global crude birth rate in 2016 was 19 per 1,000.

In 2016, crude birth rates ranged from 8 per 1,000 in countries such as Japan, Italy, Republic of Korea, and Portugal to 48 in Niger. The CBR in the United States continued trending down, as it did for the entire world since peaking in 1963, coming in at 12 per 1,000. By comparison in 1963, the world's crude birth rate hit more than 36.

Many African countries have a very high crude birth rate, and women in those countries have a high total fertility rate, meaning they give births to many children in their lifetime. Countries with a low fertility rate (and low crude birth rate of 10 to 12 in 2016) include European nations, the United States, and China.

Death Trends

The crude death rate measures the rate of deaths for every 1,000 people in a given population. Crude death rates of below 10 are considered low, while crude death rates above 20 per 1,000 are considered high. Crude death rates in 2016 ranged from 2 in Qatar, the United Arab Emirates, and Bahrain to 15 per 1,000 in Latvia, Ukraine, and Bulgaria.

The global crude death rate in 2016 was 7.6, and in the United States, the rate was 8 per 1,000. The crude death rate for the world has been on the decline since 1960 when it came in at 17.7.

It has been falling around the world (and dramatically in developing economies) due to longer life spans brought about by a better food supplies and distribution, better nutrition, better and more widely available medical care (and the development of technologies such as immunizations and antibiotics), improvements in sanitation and hygiene, and clean water supplies. Much of the increase in world population over the last century overall has been attributed more to longer life expectancies rather than an increase in births.

Migration

Migration is the movement of a person or a group of people, to settle in another place, often across a political or administrative boundary. Migration can be temporal or permanent, and it may be voluntary or forced.

There are two important terms that relate to migration:

- Immigration (people coming in from elsewhere) and Emigration (people leaving their home country). Immigration is when people move from other places into a place to settle. Such migrants are called immigrants. Emigration is when people move out to new places, and the migrants involved are called emigrants.

- Migration is not a new thing — it is known historically, that people have always had migratory lifestyles. There is enough evidence that people have moved from far away places to inhabit new areas. For example, Migrants from Asia ended up in North and South America over a period of time, via a land bridge over the Bering Strait. There has been several bulk movement of people in the history of humans, all of which were caused by some specific events during those times.

Factors of Migration

Push and pull factors in geography refer to the causes of migration among people. The reasons can be social, economic, environmental or political in nature. People migrate from a place because of unsustainable conditions such as insecurity or unemployment - these are referred to as push factors as they drive people away. The factors which attract people to live in a particular environment can include security, employment, political stability and climate. They are referred to as pull factors.

Push Factors

This refers to conditions which force people to leave their homes. A person moves because of distress. Migration is triggered by the promise of an easier and more enjoyable life elsewhere. Examples of push factors can include:

- Unemployment: Often, people leave places where they are less likely to get employment (such as rural areas) and go to urban areas where job opportunities are more plentiful. This factor has been the major reasons cities and towns are highly populated. Individuals leave their homes to search for employment in more industrialized areas.

- Insecurity: People move away from places that experience terrorism, violence, and high levels of crime. They move in search of peaceful and secure environment.

- Scarcity of land: People are forced to migrate in search of more land to cultivate and live in. Individuals in need of undertaking extensive agriculture move to less populated areas.

- Political instability: The effects of politics force people to move out of their homes or even countries, in search of a peaceful environment.

- Drought and famine: Some communities are nomads in that they move away from their land in periods of severe drought and famine in search of water and food.

Pull Factors

Pull factors refer to the factors which attract people to move to a certain area. Examples of pull factors include:

- Availability of better job opportunities: People seeking employment leave their homes to the places that they can access better opportunities.

- Religious freedom: There are places in the world where free worship is not protected. People will flee from religious prosecution.

- Political freedom: People are attracted to governments that exercise democracy as opposed to dictatorship.

- Fertile land: People interested in farming are attracted by fertile lands.

- Environmental safety: Places free from environmental hazards like flooding, earthquakes, tsunamis, and hurricanes attract a lot of people.

Demographic Transition Theory

Theory of Demographic Transition is a theory that throws light on changes in birth rate and death rate and consequently on the growth-rate of population.

Along with the economic development, tendencies of birth-rate and death rate are different. Because of it, growth rate of population is also different.

"Demographic transition refers to a population cycle that begins with a fall in the death rate, continues with a phase of rapid population growth and concludes with a decline in the birth rate"-E.G. Dolan.

According to this theory, economic development has the effect of bringing about a reduction in the death rate.

The relationship between birth and death rates changes with economic development and a country has to pass through different stages of population growth. C.P. Blacker divided population into five types as high, stationary, early expanding, low stationary and diminishing. According to the theory of demographic transition, population growth will have to pass through these different stages during the course of economic development.

The four stages of demographic transition mentioned by Max are explained as follows:

First Stage

This stage has been called high population growth potential stage. It is characterised by high and fluctuating birth and death rates which will almost neutralize each other. People mostly live in rural areas and their main occupation is agriculture which is in the stage of backwardness. The tertiary sector consisting of transport, commerce banking and insurance is underdeveloped.

All these factors are responsible for low income and poverty of the masses. Social beliefs and customs play an important role in keeping birth rate high. Death rate is also high because of primitive sanitation and absence of medical facilities. People live in dirty and unhealthy surroundings.

As a result, they are disease ridden and the absence of proper medical care results in large deaths. The mortality rate is highest among the poor. Thus, high birth rates and death rates remain approximately equal over time so that a static equilibrium with zero population growth prevails.

Second Stage

It is called the stage of Population Explosion. In this stage the death rate is decreasing while the birth rate remains constant at a high level. Agricultural and industrial productivity increases, means of transport and communication develops. There is great mobility of labour. Education expands. Income also increases. People get more and better quality of food products. Medical and health facilities are expanded.

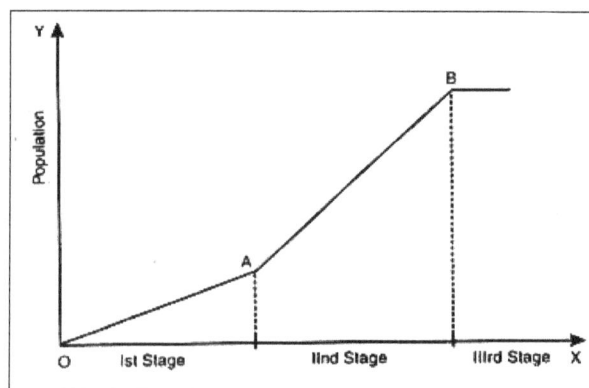

During the stage economic development is speeded up due to individual and government efforts. Increased use of better technology, mechanization and urbanisation takes place. But there is no substantial change in the men, attitude of the people and hence birth rate stays high i.e., economic development has not yet started affecting the birth rate.

Due to the widening gap between the birth and death rates, population grows at an exceptionally high rate and that is why it has been called the population explosion stage. This is an "Expanding" stage in population development where population grows at an increasing rate, as shown in figure, with the decline in death rate and no change in birth rate.

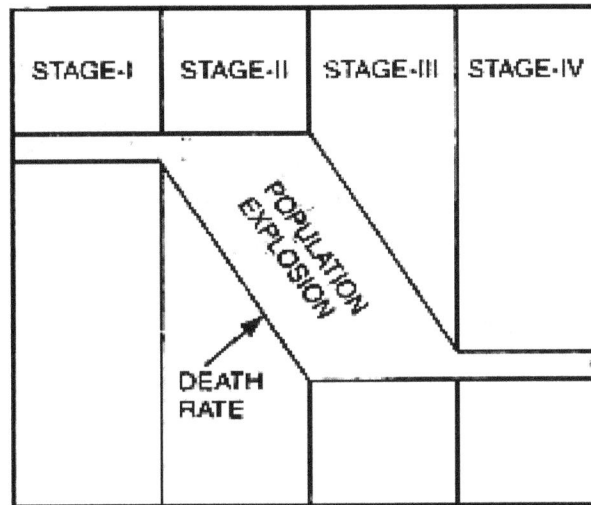

Third Stage

It is also characterised as a population stage because the population continues to grow at a fast rate. In this stage, birth rate as compared to the death rate declines more rapidly. As a result, population grows at a diminishing rate. This stage witnesses a fall in the birth rate while the death rate stays constant because it has already declined to the lowest minimum. Birth rate declines due to the impact of economic development, changed social attitudes and increased facilities for family planning. Population continues to grow fast because death rate stops falling whereas birth rate though declining but remains higher than death rate.

Fourth Stage

It is called the stage of stationary population. Birch rate and death rate are both at a low level and they are again near balance. Birth rate is approximately equal to death rate and there is little growth in population. It becomes more or less stationary at a low level.

References

- Geography: kite.kerala.gov.in, Retrieved 3 March, 2019

- Calculate-population-density: sciencing.com, Retrieved 13 May, 2019

- Factors-that-affects-population-distribution, population-geography: yourarticlelibrary.com, Retrieved 7 July, 2019

- Population-growth, human-evolution, anthropology-and-archaeology, social-sciences-and-law: encyclopedia.com, Retrieved 17 January, 2019

- Crude-birth-rate: thoughtco.com, Retrieved 14 August, 2019

- What-is-migration, migration: eschooltoday.com, Retrieved 27 June, 2019

- What-are-push-and-pull-factors: worldatlas.com, Retrieved 30 April, 2019

- Theory-of-demographic-transition, population: economicsdiscussion.net, Retrieved 6 February, 2019

Chapter 7

Human Development

The process which is involved in enlarging the freedom of people and improving their well-being is known as human development. One of the ways in which it is measured is through human development index. These diverse aspects of human development in the current scenario have been thoroughly discussed in this chapter.

Human development grew out of global discussions on the links between economic growth and development during the second half of the 20th Century. By the early 1960s there were increasingly loud calls to "dethrone" GDP: economic growth had emerged as both a leading objective, and indicator, of national progress in many countries i, even though GDP was never intended to be used as a measure of wellbeing ii. In the 1970s and 80s development debate considered using alternative focuses to go beyond GDP, including putting greater emphasis on employment, followed by redistribution with growth, and then whether people had their basic needs met.

These ideas helped pave the way for the human development approach, which is about expanding the richness of human life, rather than simply the richness of the economy in which human beings live. It is an approach that is focused on creating fair opportunities and choices for all people. So how do these ideas come together in the human development approach?

- People: the human development approach focuses on improving the lives people lead rather than assuming that economic growth will lead, automatically, to greater opportunities for all. Income growth is an important means to development, rather than an end in itself.

- Opportunities: human development is about giving people more freedom and opportunities to live lives they value. In effect this means developing people's abilities and giving them a chance to use them. For example, educating a girl would build her skills, but it is of little use if she is denied access to jobs, or does not have the skills for the local labour market. The diagram below looks at aspects of human development that are foundational (that is they are a fundamental part of human development); and aspects that are more contextual (that is they help to create the conditions that allow people to flourish). Three foundations for human development are to live a healthy and creative life, to be knowledgeable, and to have access to resources needed for a decent standard of living. Many other aspects are important too, especially in helping to create the right conditions for human development, such as environmental sustainability or equality between men and women.

- Choices: human development is, fundamentally, about more choice. It is about providing people with opportunities, not insisting that they make use of them. No one can guarantee human happiness, and the choices people make are their own concern. The process of development – human development - should at least create an environment for people, individually and collectively, to develop to their full potential and to have a reasonable chance of leading productive and creative lives that they value.

The human development approach, developed by the economist Mahbub Ul Haq, is anchored in Amartya Sen's work on human capabilities, often framed in terms of whether people are able to "be" and "do" desirable things in life. Examples include:

- Beings: well fed, sheltered, healthy

- Doings: work, education, voting, participating in community life.

Freedom of choice is central: someone choosing to be hungry (during a religious fast say) is quite different to someone who is hungry because they cannot afford to buy food.

Understanding the Difference between Growth and Development

The terms Growth and Development are used with every aspect of life. There might be some confusion when using the terms as they are often used interchangeably. Growth is just 'getting bigger', whereas development is improvement.

Growth can be explained as becoming bigger or larger or having more importance. Growth is termed as a physical change, where as development is said to be physical as well as social or psychological change. Development also means transformation or improvement. While growth is related to quantitative improvement, development is related to quantitative as well as qualitative improvement.

Pillars of Human Development

The idea of human development is supported by the concepts of equity, sustainability, productivity and empowerment.

Equity

Equity refers to creating equal access to opportunities and ensures that is available to everybody. The opportunities accessible to people must be equal irrespective of their gender, race, income.

Sustainable

Sustainability means durableness in the availability of opportunities. To have sustainable human development, each creation must have the same opportunities. All environmental, financial and human resources must be used keeping in mind the expectations. Misuse of any of these resources will lead to fewer opportunities for expected generations.

Productivity

In the case of human development productivity means human labour productivity or productivity in terms of human work. Such productivity should be constantly enriched by building capabilities in people. Eventually, it is people who are the real wealth of nations. Therefore, an effort to increase their knowledge, or provide better health facilities automatically leads to better work efficiency.

Empowerment

Empowerment means to have the power to make choices. Such power comes from increasing freedom and aptitude. Good governance and people-oriented policies are necessary to empower people. The empowerment of socially and economically underprivileged groups is of special importance.

Human Development Index

Emergence of Human Development Index

Any measure that values a gun hundred times more than a bottle of milk is bound to raise serious questions. It is no surprise, then, that since the emergence of national income accounts, there has been a considerable dissatisfaction with gross national product as a measure of human welfare. The main drawback of GNP is that it does not take into account the non-monetized activities – household work, subsistence agriculture, unpaid services. And what is more serious, GNP is one dimensional: it fails to capture the cultural, social, political and many other choices that people make.

There has been a long search for more comprehensive measure of development that could capture all, or many more, of the choices people make – a measure that would serve as a better yardstick of the socio- economic progress of nations. The search for a new composite index of socio-economic progress began in earnest in preparing the Human Development Report under the sponsorship of UNDP. Several principles guided this search.

- The new human development index (HDI) would measure the basic concept of human development to enlarge people's choices.

- The new index would include only a limited number of variables to keep it simple and manageable.

- A composite index would be constructed instead of plethora of separate indices.

- The new index would cover both social and economic choices.

- The methodology and coverage of HDI would be kept flexible – subject to gradual refinements as analytical critiques emerged and better data became available.

- Even though an index can only be as good as the data fed into it, a lack of reliable and up to date data series was not allowed to inhibit the emergence of the HDI. Instead, HDI country rankings would act as a pressure point to persuade policy makers to invest adequate amounts in producing relevant data and to encourage international institutions to prepare comparable statistical data systems.

The Human Development Indices

Even though quite a number of specific measures of HD have been presented or suggested in the literature, four of them have so far consolidated within the paradigm. These measures are the Human Development Index, the Gender-related Development Index, the Gender Empowerment Measure, and the Human Poverty Index.

a. The Human Development Index (HDI) was designed as a means to shift the emphasis from the narrow focus on economic growth (measured by GNP) to human progress and the widening of human choices, as well as to create debate on national and international policy options. HDI measures a country's total achievement in three dimensions of HD: longevity, knowledge, and a decent level of living. As variables it uses life expectancy at birth, educational achievement (literacy and combined gross schooling ratio), and the real adjusted per capita income.

b. The Human Poverty Index (HPI) measures the extent of deprivation in HDI's three dimensions. For industrialized countries, it uses as variables the probability of dying before age 60, functional illiteracy, and the incidence of poverty and long-lasting unemployment. For developing countries, its variables are the probability of death before age 40, adult illiteracy, child malnutrition, and the percentage of population with no access to drinking water.

c. The Gender-Related Development Index (GDI) measures the achievement in the three dimensions and variables of HDI, but it adjusts their values according to the inequality existing between sexes: the higher gender inequality, the larger the retrogression in the country's HDI.

d. The Gender Empowerment Measure (GEM) assesses women's participation in economic and political life. As variables it uses the female share in Parliament as well as in the higher occupational categories, and the proportion between women and men's income.

HDI, HPI and GDI refer to the same set of basic human choices (life expectancy, knowledge, and standard of living). The HD Reports and the HD academic community have explored additional dimensions, including human freedom, political democracy, inequality, poverty, technological advance, human rights, and governance. Yet an introductory course can only discuss the best known measure of HD, namely, the HDI.

Calculating the Human Development Index

As stated, three major options are chosen for HDI:

a. Long lasting and healthy life,

b. Access to knowledge, and

c. Resources for a decent life.

These options are selected for several reasons:

a. They are essential to any human life,

b. They are fairly independent from each other,

c. They cover most the spectrum of "things human beings have a good reason to value and to desire",

d. Relevant statistics are available for almost any country and for many populations of interest.

To capture the three dimensions above the following variables are used:

a. For long lasting and healthy life:

- Life expectancy at birth, precisely defined as "the average number of years a newborn can expect to live in a cohort subject to the prevailing age-specific mortality rates in the society and moment under consideration".

b. For access to knowledge:

- Adult (over age 15) literacy rate, where literacy is understood as "the capability of reading, writing and understanding a simple and short text on everyday life"; and

- Combined gross enrollment ratio for population aged 6 to 23, where the combined gross enrollment ratio is the total number of students enrolled in primary, secondary or tertiary formal education divided by the total population in the corresponding ages.

c. For resources for a decent life:

The per capita income expressed both in US dollars and in purchasing power parity - PPP-units.

- The per capita income is the total value of the final goods and services produced by a country in a given period, divided by the total population at mid-year.

- The per capita income, usually expressed in nominal US dollars, fails to consider the inter-country variations in the cost of living. Units of purchasing power parity (PPP) are used to correct this deficiency. The correction is based on calculating the international or average price of a large series of goods and services in different countries, and in applying those standard prices to the goods and services of the country in question.

To choosing these variables, conceptual problems do arise. In fact, it can be questioned whether or not life expectancy is the best measure of a long lasting and healthy life; whether or not access to knowledge could be measured with variables better than literacy and schooling, or if the resources for a decent life could be better measured with indicators other than the per capita income in PPP.

The several goals of HD "cannot be reduced to a single variable or merely to a number"; and "the range of HD choices is, in principle, endless". Both statements seemingly run in the face of the HDI, which pins HD down to a number and is based in but three human choices.

The concept of HD is much broader than the three dimensions included in the HDI. For example, the HDI does not reflect political participation, governance issues or gender inequalities. This is largely because of the difficulty of adequately capturing such complex aspects of HD in a single index, and due to the absence of some generally agreed and unambiguous indicators. A fuller picture of a country's level of HD requires an analysis of other HD indicators and information as well.

HDI cannot be used as a measure of HD change in the short-run, as the effect of policies to impact two of the HDI indicators-adult literacy and life expectancy-will only be felt long after having put these policies in place. As a result of this lag time, the HDI best captures long-run changes in a country's HD situation.

Rather than a paralyzing criticism, the above points should be taken as cautionary notes in using and interpreting the HDI (and the remaining measures of HD). For one thing, these indices do not claim to reflect the full range of HD choices; they select a few yet highly relevant choices. Then, they do not aim at measuring the "true" development of any one country but, instead, at ranking countries according to their HD status. Lastly, in their coming out but with a number, the indices are not meant to ignore HD's multidimensionality; they simply improve upon standard, unidimensional measures of development.

Value Addition of Human Development Index

The HDI is a measure that can capture the attention of policy makers, media and NGOs and expand the debate beyond the more usual economic statistics to focus instead on human outcomes.

The HDI can also provide a basis for questioning national policy choices:

- The Philippines 1999 report on education spurred debates on educational reforms in the country's Senate and Executive Cabinet, and the 1997 report led to Presidential directive mandating all local governments to devote at least 20 per cent of the revenue to HD priorities. The President also asked the National Statistical Coordination Board to include the Human Development Index (HDI) in the system of statistics to track variations across provinces.

- Japan and South Korea have adopted the HDR's Gender Empowerment Measure in the formulation of national legislation.

- The HDI can also highlight wide differences within countries, between provinces or states, across gender, ethnicity, and other socioeconomic groupings. Highlighting internal disparities along these lines has raised national debate in many countries.

- In the Egypt HDR 1999, HDI revealed that Upper Egypt region was far behind Cairo in every dimension of human development. This led to a formal policy discussion of resource allocation between the governors of 17 provinces in the country and the entire resource allocation pattern was changed to funnel more funds to the Upper Egypt region.

- In Brazil, the large State of Minas Gerais, disaggregated the human development index for all its municipalities. It then introduced the so called "Robin Hood Law" that ensures that more tax revenues are allocated to those of its municipalities that rank low on the index, as well as perform poorly on a number of other social and environmental indicators. The central government is now planning to use a modified version of the human development index, in combination with other indicators, to allocate resources to all of the country's more than 5,000 municipalities. No longer will population size be used as the only criteria when resource allocations to municipalities are determined. Instead the budgets will depend on their level of human development.

Human Development and Human Development Index

Ironically, the human development approach to development has fallen victim to the success of human development index. The HDI has reinforced the narrow, oversimplified interpretation of the human development concept as being only about expanding education, health and decent living standards. This had obscured the broader, more complex concept of human development as the expansion of capabilities that widen the peoples choices to lead lives that they value.

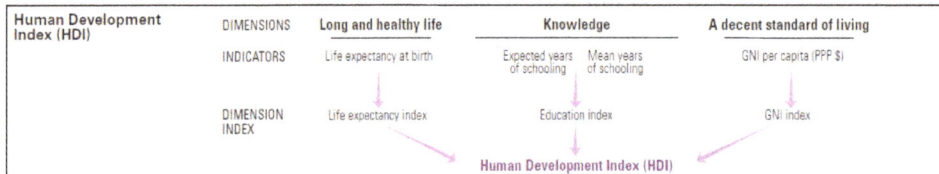

Despite careful efforts to explain that the concept is broader than the measure, human development continues to be identified with the HDI- while political freedoms, participating in the life of one's community and physical security are often overlooked. But such capabilities are as universal and fundamental as being able to read or to enjoy good health. They are valued by all people- and without them all choices are foreclosed. They are not included in the HDI because they are difficult to measure appropriately, not because they are less important to human development.

Always remember that HDI is just a summary measure and does not provide a comprehensive picture of human development in any situation.

References

- What-human-development: hdr.undp.org, Retrieved 9 August, 2019

- Difference-between-growth-and-development, language: differencebetween.net, Retrieved 19 April, 2019

- What-is-human-development-analyse-the-four-pillars-of-human: school.gradeup.co, Retrieved 23 February, 2019

- Measuring-human-development: undp.org, Retrieved 3 May, 2019

- Human-development-index: undp.org, Retrieved 13 March, 2019

Chapter 8

Understanding Oceanography

The branch of science which is involved in the study of all the aspects of ocean is known as oceanography. Some of the areas of study within this discipline are ocean temperature, ocean salinity and ocean floor relief. The chapter closely examines these key aspects of oceanography to provide an extensive understanding of the subject.

Hydrological Cycle

The Hydrologic Cycle is one of the most important processes in the natural world, and is perhaps one that we all take for granted. All of the world's water is subject to this process, which sees the water change forms, locations, and accessibility.

In its most basic assessment, water changes between three different states in this cycle. It variously takes the form of liquid, gas, and solid: water, steam or vapor, and ice. Throughout the cycle, the water will undergo changes between these three forms many times: water freezing into ice, ice melting into water, water evaporating into water vapor, and that vapor then condensing to become water once more.

The hydrologic cycle, also known as global water cycle or the H_2O cycle, describes the storage and movement of water between the biosphere, atmosphere, lithosphere, and the hydrosphere. "The water cycle, also known as the hydrological cycle or the H_2O cycle, describes the continuous movement of water on, above and below the surface of the Earth."

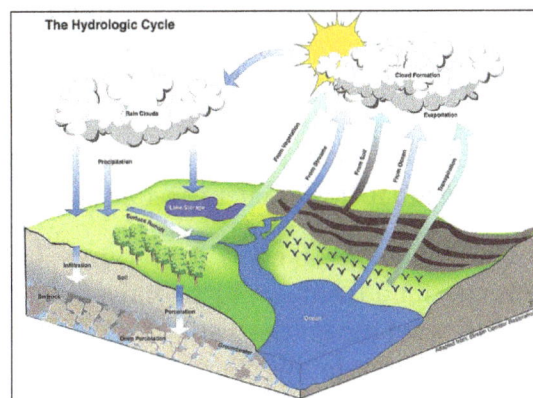

Water is most commonly found in its liquid form, in rivers, oceans, streams, and in the earth. The sun's rays constantly warm the water found in these places and, whether through this heat or through man-made means, the water particles gain energy and spread, turning the water from a liquid into a vapor through evaporation. The water vapor, thus becoming less dense, rises with the warm air into the sky where it sticks to other water particles to form clouds.

Typically, we consider the boiling point of water to be a hundred degrees centigrade, which is certainly true when pressure and humidity are normal. However, places such as mountains, where humidity is low and pressure is even lower, require less energy to boil away the water.

Along with the water vapor, some small particles can often rise up to form clouds. It is not only liquid water that can evaporate to become water vapor, but ice and snow, too. This process is simple enough, however there are a few things to note about evaporation.

This simple explanation, however, does not do justice to the complexity of the hydrologic cycle, which comprises many more steps.

Different Steps of the Hydrologic Cycle

Here is a breakdown of the different steps of the hydrologic cycle.

Water is most commonly found in its liquid form, in rivers, oceans, streams, and in the earth. The sun's rays constantly warm the water found in these places and, whether through this heat or through man-made means, the water particles gain energy and spread, turning the water from a liquid into a vapor through evaporation. The water vapor, thus becoming less dense, rises with the warm air into the sky where it sticks to other water particles to form clouds.

- Evaporation – is frequently used as a catch-all term to refer to the process of water turning to water vapor, however there is another distinct term for the evaporation of water from a plant's leaves.

- Evapotranspiration – makes up a large portion of the water in the planet's atmosphere due to the sheer surface area of the globe covered by flora. The majority of water in the atmosphere comes from lakes and oceans – around ninety per cent – but in terms of land-based water, evapotranspiration is an important player.

- Sublimation – as the process is called, results from when pressure and humidity are low as noted above. It is not only liquid water that can evaporate to become water vapor, but ice and snow, too. Due to lower air pressure, less energy is required to sublimate the ice into vapor. Other factors which can aid in sublimation are high winds and strong sunlight, which is why mountain ice is a prime candidate for sublimation, while ground ice sublimation is not so common. A good, visible example of sublimation is dry ice, which emits a thick layer of water vapor due to its lower energy requirement.

  When water vapor reaches this plane, it cools significantly and clumps together. So stuck together, this newly formed cloud is subject to the movement of the wind and the changes in the air pressure, which is what moves the water around the planet. There are a couple of things that can happen to the vapor in this state.

- Precipitation/Rainfall – refers to vapor that cools to any temperature above freezing point (zero degrees centigrade) will condense, becoming droplets of liquid water. These droplets form when the water vapor condenses around particles and other matter that rises up with the water during evaporation, giving a nucleus to the water droplet so that it can clump together. Once a number of these tiny, particle based droplets form, they collide and clump together as larger droplets. At a certain point, the droplet will become big enough that its mass will be subject to the force of gravity at a rate faster than the force of the updraft in the air around it. At this point, the water falls to earth.

- Snow – refers to frozen water falling from the sky. When it is particularly cold or the air pressure is exceptionally low, these water droplets will crystalize before falling.

- Sleet – is a bitterly cold, half-frozen slush. This third state occurs when the conditions are not quite cold enough to keep the crystals frozen and the water either does not freeze fully or if precipitation occurs in particularly cold conditions, or conditions in which the air pressure is very low, then these water droplets can quite often crystallize and freeze. This causes the water to fall as solid ice, known melts somewhat in the process.

When water falls to earth, it quite often ends up on tarmac or over man-made surfaces where it quickly evaporates again.

- Infiltration – is water that doesn't evaporate after precipitation and falls into soil and other absorbent surfaces. The water moves throughout the soil, saturating it.

- Groundwater Storage – is water that has not precipitated or run off into streams or rivers, but instead moves deep underground forming pools known as "groundwater storage". In groundwater storage, water joins up in the soil and forms pools of saturated soil instead of escaping the soil. These pools are called "aquifers".

- Springs – occur when an aquifer becomes oversaturated, and the excess water leaks out of the soil onto the surface. Most commonly, springs will emerge from cracks in rocks and holes in the ground. Sometimes, if conditions are particularly volcanic, the spring will heat up and form "hot springs".

- Runoff – After heavy rainfall has saturated the soil it will cease to absorb water and additional rainfall, as well as melted snow and ice, will simply flow off of the surface. The flow follows gravity down hills, mountains, and other inclines to form streams and join rivers. This is known as "runoff", and it is the principle way in which water moves along the Earth's surface. The rivers and streams are pulled by gravity until they pool together to form lakes and oceans.

- Streamflow – is the direction the runoff takes to form a stream and it is this flow which dictates the river's currents depending on how close they are to the ocean. Because ice and snow make up a large portion of the water involved in runoff, heatwaves are a principle cause of flooding as the water stored on the surface is suddenly released into runoff flow. In particular, a warm spring following a cold winter can result in quite spectacular flood, as a large volume of water gets stored in ice and snow only to quickly melt and form new streams.

- Ice Caps – occur when a large volume of snow falls and is not evaporated or sublimated, the ice compacts under its own weight to form these caps. Ice caps, glaciers, and ice sheets contain a huge amount of water, and those found in the polar regions of the planet are the largest stores of ice found in the world. As the atmosphere warms up slowly, more and more of this ice melts and evaporates, releasing more water into the hydrologic cycle. It is this process which causes rises in the ocean levels.

The hydrologic cycle happens continuously, with all different steps happening simultaneously around the world. The biggest concern that many have with the hydrologic cycle is the availability of drinkable water, which is something that is constantly in flux, and the melting of the huge ice storage sheets at the polar caps. Having an understanding of the different steps of the hydrologic cycle is an important step in understanding what effect human activity has on the world's water.

Ocean Floor Relief

Ocean Relief

- Ocean relief is largely due to tectonic, volcanic, erosional and depositional processes and their interactions.
- Ocean relief features are divided into major and minor relief features.

Major Ocean Relief Features

Four major divisions in the ocean relief are:

1. The continental shelf,
2. The continental slope,
3. The continental rise,
4. The Deep Sea Plain or the abyssal plain.

Minor Ocean Relief Features

- Ridges,
- Hills,
- Seamounts,
- Guyots,
- Trenches,
- Canyons,
- Sleeps,
- Fracture zones,
- Island arcs,
- Atolls,
- Coral reefs,
- Submerged volcanoes and

- Sea-scarps.

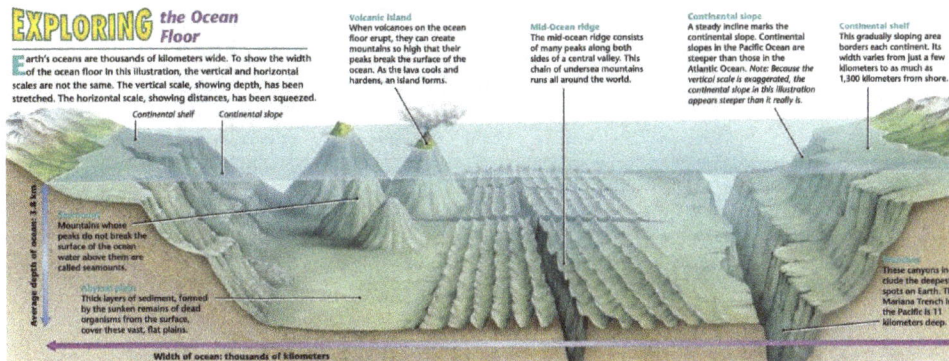

Continental Shelf

- Continental Shelf is the gently sloping seaward extension of continental plate.

- These extended margins of each continent are occupied by relatively shallow seas and gulfs.

- Continental Shelf of all oceans together cover 7.5% of the total area of the oceans.

- Gradient of continental is of 1° or even less.

- The shelf typically ends at a very steep slope, called the shelf break.

- The continental shelves are covered with variable thicknesses of sediments brought down by rivers, glaciers etc.

- Massive sedimentary deposits received over a long time by the continental shelves, become the source of fossil fuels [Petroleum].

- Examples: Continental Shelf of South-East Asia, Great Banks around Newfoundland, Submerged region between Australia and New Guinea.

- The shelf is formed mainly due to-

  ○ Submergence of a part of a continent,

  ○ Relative rise in sea level,

  ○ Sedimentary deposits brought down by rivers.

- There are various types of shelves based on different sediments of terrestrial origin-

  ○ Glaciated shelf (Surrounding Greenland),

  ○ Coral reef shelf (Queensland, Australia),

  ○ Shelf of a large river (Around Nile Delta),

  ○ Shelf with dendritic valleys (At the Mouth of Hudson River),

    ◦   Shelf along young mountain ranges (Shelves between Hawaiian Islands).

Width

- The average width of continental shelves is between 70 – 80 km.

- The shelves are almost absent or very narrow along some of the margins like the coasts of Chile, the west coast of Sumatra, etc. [Ocean – Continent Convergence and Ocean – Ocean Convergence].

- It is up to 120 km wide along the eastern coast of USA. On the contrary, the Siberian shelf in the Arctic Ocean, the largest in the world, stretches to 1,500 km in width.

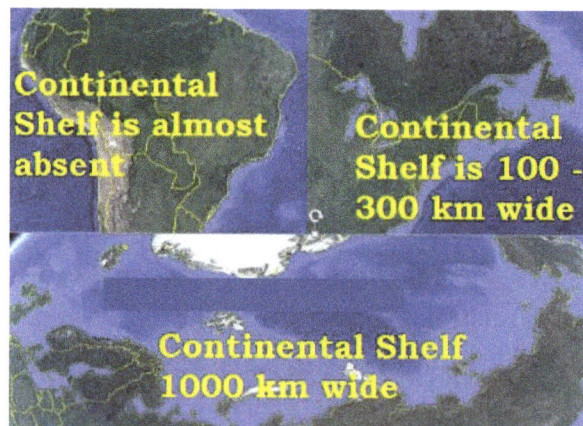

Depth

- The depth of the shelves also varies. It may be as shallow as 30 m in some areas while in some areas it is as deep as 600 m.

Importance of continent shelves:

1. Marine food comes almost entirely from continental shelves;

2. They provide the richest fishing grounds;

3. They are potential sites for economic minerals 20% of the world production of petroleum

and gas comes from shelves. Polymetallic nodules (manganese nodules; concentric layers of iron and manganese hydroxides) etc. are good sources of various mineral ores like manganese, iron copper, gold etc.

Continental Slope

- The continental slope connects the continental shelf and the ocean basins.

- It begins where the bottom of the continental shelf sharply drops off into a steep slope.

- The gradient of the slope region varies between 2-5°.

- The depth of the slope region varies between 200 and 3,000 m.

- The seaward edge of the continental slope loses gradient at this depth and gives rise to continental rise.

- The continental slope boundary indicates the end of the continents.

- Canyons and trenches are observed in this region.

Continental Rise

- The continental slope gradually loses its steepness with depth.

- When the slope reaches a level of between 0.5° and 1°, it is referred to as the continental rise.

- With increasing depth the rise becomes virtually flat and merges with the abyssal plain.

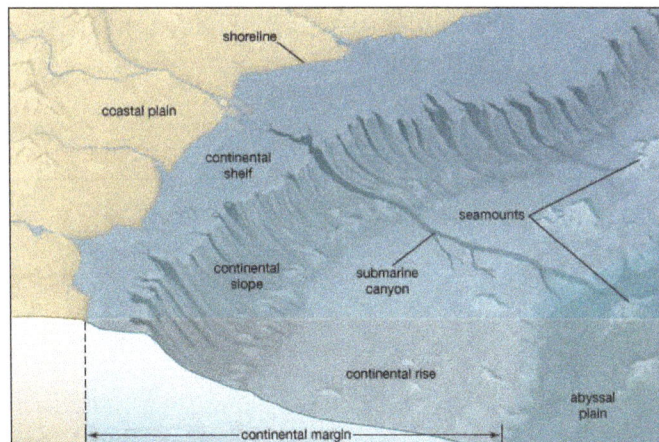

Deep Sea Plain or Abyssal Plain

- Deep sea planes are gently sloping areas of the ocean basins.

- These are the flattest and smoothest regions of the world because of terrigenous [denoting marine sediment eroded from the land] and shallow water sediments that buries the irregular topography.

- It covers nearly 40% of the ocean floor.

- The depths vary between 3,000 and 6,000 m.

- These plains are covered with fine-grained sediments like clay and silt.

Oceanic Deeps or Trenches

- The trenches are relatively steep sided, narrow basins (Depressions). These areas are the deepest parts of the oceans.

- They are of tectonic origin and are formed during ocean – ocean convergence and ocean continent convergence.

- They are some 3-5 km deeper than the surrounding ocean floor.

- The trenches lie along the fringes of the deep-sea plain at the bases of continental slopes and along island arcs.

- The trenches run parallel to the bordering fold mountains or the island chains.

- The trenches are very common in the Pacific Ocean and form an almost continuous ring along the western and eastern margins of the Pacific.

- The Mariana Trench off the Guam Islands in the Pacific Ocean is the deepest trench with, a depth of more than 11 kilometres.

- They are associated with active volcanoes and strong earthquakes (Deep Focus Earthquakes like in Japan). This makes them very significant in the study of plate movements.

- As many as 57 deeps have been explored so far; of which 32 are in the Pacific Ocean; 19 in the Atlantic Ocean and 6 in the Indian Ocean.

Mid-Oceanic Ridges or Submarine Ridges

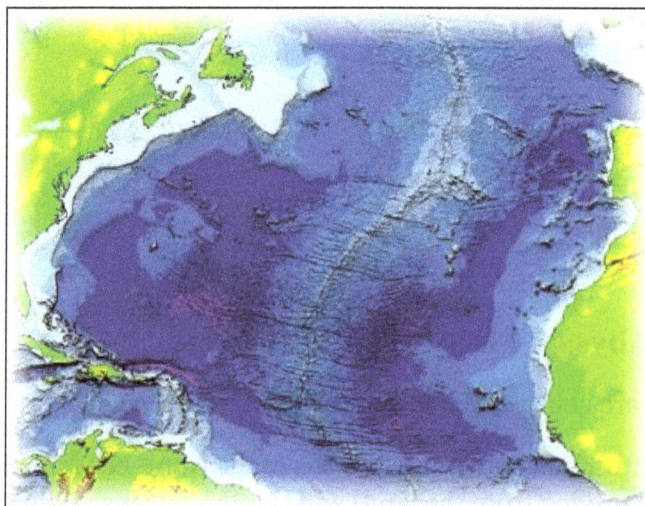

- A mid-oceanic ridge is composed of two chains of mountains separated by a large depression.

- The mountain ranges can have peaks as high as 2,500 m and some even reach above the ocean's surface.

- Running for a total length of 75,000 km, these ridges form the largest mountain systems on earth.

- These ridges are either broad, like a plateau, gently sloping or in the form of steep-sided narrow mountains.

- These oceanic ridge systems are of tectonic origin and provide evidence in support of the theory of Plate Tectonics.

- Iceland, a part of the mid-Atlantic Ridge, is an example.

Abyssal Hills

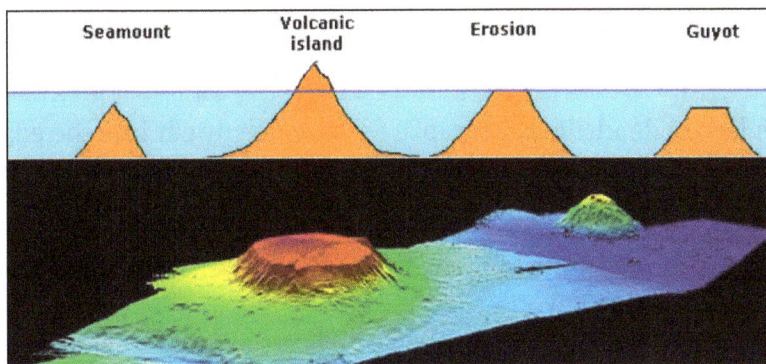

- Seamount: It is a mountain with pointed summits, rising from the seafloor that does not reach the surface of the ocean. Seamounts are volcanic in origin. These can be 3,000-4,500 m tall.

- The Emperor seamount, an extension of the Hawaiian Islands [Hotspot] in the Pacific Ocean, is a good example.

- Guyots: The flat topped mountains (seamounts) are known as guyots.

- Seamounts and guyots are very common in the Pacific Ocean where they are estimated to number around 10,000.

Submarine Canyons

- Canyon: a deep gorge, especially one with a river flowing through it.

- Gorge: a steep, narrow valley or ravine.

- Valley: a low area between hills or mountains or a depression, typically with a river or stream flowing through it.

- These are deep valleys, some comparable to the Grand Canyon of the Colorado river.

- They are sometimes found cutting across the continental shelves and slopes, often extending from the mouths of large rivers.

- The Hudson Canyon is the best known canyon in the world.

Broadly, there are three types of submarine canyons:

- Small gorges which begin at the edge of the continental shelf and extend down the slope to very great depths, e.g., Oceanographer Canyons near New England.

  Those which begin at the mouth of a river and extend over the shelf, such as the Zaire, the Mississippi and the Indus canyons.

- Those which have a dendritic appearance and are deeply cut into the edge of the shelf and the slope, like the canyons off the coast of southern California. The Hudson Canyon is the best known canyon in the world.

- The largest canyons in the world occur in the Bering Sea off Alaska. They are the Bering, Pribilof and Zhemchung canyons.

Atoll

- These are low islands found in the tropical oceans consisting of coral reefs surrounding a central depression.

- It may be a part of the sea (lagoon), or sometimes form enclosing a body of fresh, brackish, or highly saline water.

Bank, Shoal and Reef

- These marine features are formed as a result of erosional, depositional and biological activity.

- These are produced upon features of diastrophic [Earth Movements] origin. Therefore, they are located on upper parts of elevations.

Bank

- These marine features are formed as' a result of erosional and depositional activity.

- A bank is a flat topped elevation located in the continental margins.

- The depth of water here is shallow but enough for navigational purposes.

- The Dogger Bank in the North Sea and Grand Bank in the north-western Atlantic, Newfoundland are famous examples.

- The banks are sites of some of the most productive fisheries of the world.

Shoal

- A shoal is a detached elevation with shallow depths. Since they project out of water with moderate heights, they are dangerous for navigation.

Reef

- A reef is a predominantly organic deposit made by living or dead organisms that forms a mound or rocky elevation like a ridge.

- Coral reefs are a characteristic feature of the Pacific Ocean where they are associated with seamounts and guyots.

- The largest reef in the world is found off the Queensland coast of Australia.

- Since the reefs may extend above the surface, they are generally dangerous for navigation.

Ocean Temperature

Factors affecting Temperature of Ocean Water

The following factors affect the distribution of temperature of ocean water:

Latitudes

The temperature of surface water decreases from equator towards the poles because the sun's rays become more and more slanting and thus the amount of insolation decreases poleward accordingly. The temperature of surface water between 40 °N and 40 °S is lower than air temperature but it becomes higher than air temperature between 40th latitude and the poles in both the hemispheres.

Unequal Distribution of Land and Water

The temperature of ocean water varies in the northern and the southern hemispheres because of dominance of land in the former and water in the latter. The oceans in the northern hemisphere receive more heat due to their contact with larger extent of land than their counterparts in the southern hemisphere and thus the temperature of surface water is comparatively higher in the former than the latter.

The isotherms are not regular and do not follow latitudes in the northern hemisphere because of the existence of both warm and cold land- masses whereas they (isotherms) are regular and follow latitudes in the southern hemisphere because of the dominance of water. The temperature in the enclosed seas in low latitudes becomes higher because of the influence of surrounding land areas than the open seas e.g., the average annual temperature of surface water at the equator is 26.7 °C (80 °F) whereas it is 37.8 °C (100 °F) in the Red Sea and 34.4 °C (94 °F) in the Persian Gulf.

Prevailing Wind

Wind direction largely affects the distribution of temperature of ocean water. The winds blowing from the land towards the oceans and seas (e.g., offshore winds) drive warm surface water away from the coast resulting into upwelling of cold bottom water from below. Thus, the replacement of warm water by cold water introduces longitudinal variation in temperature. Contrary to this, the onshore winds pile up warm water near the coast and thus raise the temperature.

For example, trade winds cause low temperature (in the tropics along the eastern margins of the oceans or the western coastal regions of the continents) because they blow from the land towards the oceans whereas these trade winds raise the temperature in the western margins of the oceans or the eastern coastal areas of the continents because of their onshore position.

Similarly, the eastern margins of the oceans in the middle latitudes (western coasts of Europe and North America) have relatively higher temperature than the western margins of the oceans because of the onshore position of the westerlies.

Ocean Currents

Surface temperatures of the oceans are controlled by warm and cold currents. Warm currents raise the temperature of the affected areas whereas cool currents lower down the temperature. For example, the Gulf Stream raises the temperature near the eastern coasts of N. America and the western coasts of Europe.

Kuro Shivo drives warm water away from the eastern coast of Asia and raises the temperature near Alaska. Labrador cool current lowers down the temperature near north-east coast of N. America. Similarly, the temperature of the eastern coast of Siberia becomes low due to Kurile cool current.

It may be mentioned that warm currents raise the temperature more in the northern hemisphere than in the southern hemisphere which is apparent from the fact that the 5°C isotherm reaches 70° latitude in the northern Atlantic Ocean whereas it is extended up to only 50° latitude in the southern Atlantic Ocean. This is because of more dominant effects of the warm Brazil current in the southern Atlantic Ocean.

Minor Factors

Minor factors include:

- Submarine ridges,

- Local weather conditions like storms, cyclones, hurricanes, fog, cloudiness, evaporation and condensation, and

- Location and shape of the sea.

Longitudinally more extensive seas in the low latitudes have higher temperature than the latitudinally more extensive seas as the Mediterranean Sea records higher temperature than the Gulf of California.

The enclosed seas in the low latitudes record relatively higher temperature than the open seas whereas the enclosed seas have lower temperature than the open seas in the high latitudes (Baltic Sea records 0 °C (32 °F) and open seas have 4.4 °C or 40 °F).

Horizontal and Vertical Distribution of Temperature

The study of the temperature of the oceans is important for determining the,

- Movement of large volumes of water (vertical and horizontal ocean currents),

- Type and distribution of marine organisms at various depths of oceans,

- Climate of coastal lands, etc.

Source of Heat in Oceans

- The sun is the principal source of energy (Insolation).

- The ocean is also heated by the inner heat of the ocean itself (earth's interior is hot. At the sea surface, the crust is only about 5 to 30 km thick). But this heat is negligible compared to that received from sun.

Deep water marine organisms survive in absence of sunlight:

- Photic zone is only about few hundred meters. It depends on lot of factors like turbidity, presence of algae etc.

- There are no enough primary producers below few hundred meters till the ocean bottom.

- At the sea bottom, there are bacteria that make use of heat supplied by earth's interior to prepare food. So, they are the primary producers.

- Other organisms feed on these primary producers and subsequent secondary producers.

- So, the heat from earth supports wide ranging deep water marine organisms.

- But the productivity is too low compared to ocean surface.

Why is diurnal range of ocean temperatures too small? Why oceans take more time to heat or cool?

- The process of heating and cooling of the oceanic water is slower than land due to vertical and horizontal mixing and high specific heat of water.

- (More time required to heat up a Kg of water compared to heating the same unit of a solid at same temperatures and with equal energy supply).

The ocean water is heated by three processes:

- Absorption of sun's radiation.

- The conventional currents: Since the temperature of the earth increases with increasing depth, the ocean water at great depths is heated faster than the upper water layers. So, convectional oceanic circulations develop causing circulation of heat in water.

- Heat is produced due to friction caused by the surface wind and the tidal currents which increase stress on the water body.

The ocean water is cooled by:

- Back radiation (heat budget) from the sea surface takes place as the solar energy once received is reradiated as long wave radiation (terrestrial radiation or infrared radiation) from the seawater.

- Exchange of heat between the sea and the atmosphere if there is temperature difference.

- Evaporation: Heat is lost in the form of latent heat of evaporation (atmosphere gains this heat in the form of latent heat of condensation).

Factors affecting Temperature Distribution of Oceans

- Insolation: The average daily duration of insolation and its intensity.

- Heat loss: The loss of energy by reflection, scattering, evaporation and radiation.

- Albedo: The albedo of the sea (depending on the angle of sun rays).

- The physical characteristics of the sea surface: Boiling point of the sea water is increased in the case of higher salinity and vice versa [Salinity increased == Boiling point increased == Evaporation decreased].

- The presence of submarine ridges and sills [Marginal Seas]: Temperature is affected due to lesser mixing of waters on the opposite sides of the ridges or sills.

- The shape of the ocean: The latitudinally extensive seas in low latitude regions have warmer surface water than longitudinally extensive sea [Mediterranean Sea records higher temperature than the longitudinally extensive Gulf of California].

- The enclosed seas (Marginal Seas – Gulf, Bay etc.) in the low latitudes record relatively higher temperature than the open seas; whereas the enclosed seas in the high latitudes have lower temperature than the open seas.

- Local weather conditions such as cyclones.

- Unequal distribution of land and water: The oceans in the northern hemisphere receive more heat due to their contact with larger extent of land than the oceans in the southern hemisphere.

- Prevalent winds generate horizontal and sometimes vertical ocean currents: The winds blowing from the land towards the oceans (off-shore winds-moving away from the shore) drive warm surface water away from the coast resulting in the upwelling of cold water from below (This happens near Peruvian Coast in normal years. El-Nino).

- Contrary to this, the onshore winds (winds flowing from oceans into continents) pile up warm water near the coast and this raises the temperature (This happens near the Peruvian coast during El Nino event)(In normal years, North-eastern Australia and Western Indonesian islands see this kind of warm ocean waters due to Walker Cell or Walker Circulation).

- Ocean currents: Warm ocean currents raise the temperature in cold areas while the cold currents decrease the temperature in warm ocean areas. Gulf stream (warm current) raises the temperature near the eastern coast of North America and the West Coast of Europe while the Labrador current (cold current) lowers the temperature near the north-east coast of North America (Near Newfoundland). All these factors influence the temperature of the ocean currents locally.

Vertical Temperature Distribution of Oceans

- Photic or euphotic zone extends from the upper surface to ~200 m. The photic zone receives adequate solar insolation.

- Aphotic zone extends from 200 m to the ocean bottom; this zone does not receive adequate sunrays.

Thermocline

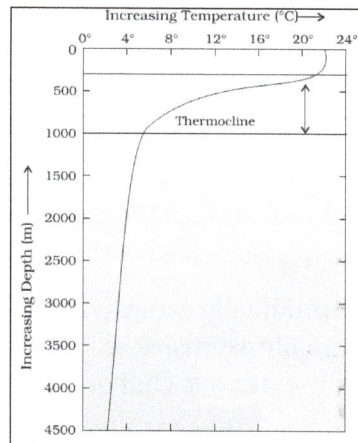

Thermocline.

- The profile shows a boundary region between the surface waters of the ocean and the deeper layers.

- The boundary usually begins around 100 – 400 m below the sea surface and extends several hundred of meters downward.

- This boundary region, from where there is a rapid decrease of temperature, is called the thermocline. About 90 per cent of the total volume of water is found below the thermocline in the deep ocean. In this zone, temperatures approach 0° C.

Three-Layer System

- The temperature structure of oceans over middle and low latitudes can be described as a three-layer system from surface to the bottom.

- The first layer represents the top layer of warm oceanic water and it is about 500m thick with temperatures ranging between 20° and 25° C. This layer, within the tropical region, is present throughout the year but in mid-latitudes it develops only during summer.

- The second layer called the thermocline layer lies below the first layer and is characterized by rapid decrease in temperature with increasing depth. The thermocline is 500 -1,000 m thick.

- The third layer is very cold and extends up to the deep ocean floor. Here the temperature becomes almost stagnant.

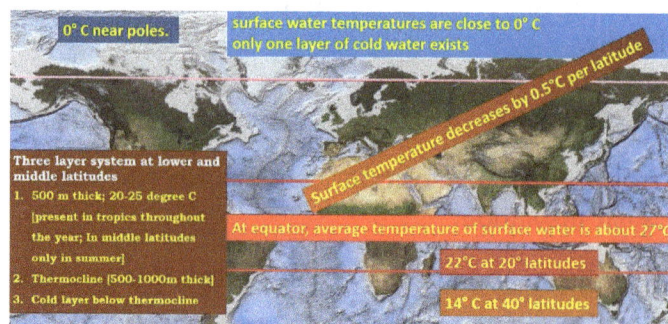

General Behavior

- In the Arctic and Antarctic circles, the surface water temperatures are close to 0 °C and so the temperature change with the depth is very slight (ice is a very bad conductor of heat). Here, only one layer of cold water exists, which extends from surface to deep ocean floor.

- The rate of decrease of temperature with depths is greater at the equator than at the poles.

- The surface temperature and its downward decrease is influenced by the upwelling of bottom water (Near Peruvian coast during normal years).

- In cold Arctic and Antarctic regions, sinking of cold water and its movement towards lower latitudes is observed.

- In equatorial regions the surface, water sometimes exhibits lower temperature and salinity due to high rainfall, whereas the layers below it have higher temperatures.

- The enclosed seas in both the lower and higher latitudes record higher temperatures at the bottom.

- The enclosed seas of low latitudes like the Sargasso Sea, the Red Sea and the Mediterranean Sea have high bottom temperatures due to high insolation throughout the year and lesser mixing of the warm and cold' waters.

- In the case of the high latitude enclosed seas, the bottom layers of water are warmer as water of slightly higher salinity and temperature moves from outer ocean as a sub-surface current.

- The presence of submarine barriers may lead to different temperature conditions on the two sides of the barrier. For example, at the Strait of Bab-el-Mandeb, the submarine barrier (sill) has a height of about 366 m. The subsurface water in the strait is at high temperature compared to water at same level in Indian ocean. The temperature difference is greater than nearly 20 °C.

Horizontal Temperature Distribution of Oceans

- The average temperature of surface water of the oceans is about 27 °C and it gradually decreases from the equator towards the poles.

- The rate of decrease of temperature with increasing latitude is generally 0.5 °C per latitude.

- The horizontal temperature distribution is shown by isothermal lines, i.e., lines joining places of equal temperature.

- Isotherms are closely spaced when the temperature difference is high and vice versa.

- For example, in February, isothermal lines are closely spaced in the south of Newfoundland, near the west coast of Europe and North Sea and then isotherms widen out to make; a bulge towards north near the coast of Norway. The cause of this phenomenon lies in the cold Labrador Current flowing southward along the north American coast which reduces the temperature of the region more sharply than in other places in the same latitude; at the same time the warm Gulf Stream proceeds towards the western coast of Europe and raises the temperature of the west coast of Europe.

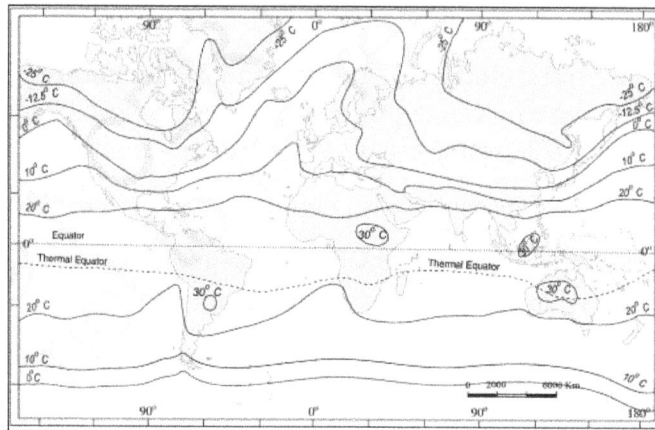

Range of Ocean Temperature

- The oceans and seas get heated and cooled slower than the land surfaces. Therefore, even if the solar insolation is maximum at noon, the ocean surface temperature is highest at 2 p.m.

- The average diurnal or daily range of temperature is barely 1 degree in oceans and seas.

- The highest temperature in surface water is attained at 2 p.m. and the lowest, at 5 a.m.

- The diurnal range of temperature is highest in oceans if the sky is free of clouds and the atmosphere is calm.

- The annual range of temperature is influenced by the annual variation of insolation, the nature of ocean currents and the prevailing winds.

- The maximum and the minimum temperatures in oceans are slightly delayed than those of land areas the maximum being in August and the minimum in February [Think why intense tropical cyclones occur mostly between August and October – case is slightly different in Indian Ocean due to its shape].

- The northern Pacific and northern Atlantic oceans have a greater range of temperature than their southern parts due to a difference in the force of prevailing winds from the land and more extensive ocean currents in the southern parts of oceans.

- Besides annual and diurnal ranges of temperature, there are periodic fluctuations of sea temperature also. For example, the 11-year sunspot cycle causes sea temperatures to rise after a 11- year gap.

Ocean Salinity

- Salinity is the term used to define the total content of dissolved salts in sea water.

- It is calculated as the amount of salt (in gm) dissolved in 1,000 gm (1 kg) of seawater.

- It is usually expressed as parts per thousand or ppt.

- Salinity of 24.7 (24.7 o/oo) has been considered as the upper limit to demarcate 'brackish water'.

Role of Ocean Salinity

- Salinity determines compressibility, thermal expansion, temperature, density, absorption of insolation, evaporation and humidity.

- It also influences the composition and movement of the sea: water and the distribution of fish and other marine resources.

Highest salinity in water bodies Lake Van in Turkey ($330\ °/_{oo}$), Dead Sea ($238°/_{oo}$ V.), Great Salt Lake ($220°/_{oo}$).

Dissolved salts in sea water(gm of salt per kg of water)	
Chlorine	18.97
Sodium	10.47
Sulphate	2.65
Magnesium	1.28
Calcium	0.41
Potassium	0.38
Bicarbonate	0.14
Bromine	0.06
Borate	0.02
Strontium	0.01

Factors affecting Ocean Salinity

- Surface salinity is greatly influenced in coastal regions by the fresh water flow from rivers, and in polar regions by the processes of freezing and thawing of ice.

- Wind, also influences salinity of an area by transferring water to other areas.

- The ocean currents contribute to the salinity variations.

- Salinity, temperature and density of water are interrelated. Hence, any change in the temperature or density influences the salinity of an area.

Horizontal Distribution of Salinity

To make life easier, you will be remove the symbol o/oo and can place only number.

- The salinity for normal open ocean ranges between 33 and 37.

High Salinity Regions

- In the land locked Red Sea (don't confuse this to Dead Sea which has much greater salinity), it is as high as 41.

- In hot and dry regions, where evaporation is high, the salinity sometimes reaches to 70.

Comparatively Low Salinity Regions

- In the estuaries (enclosed mouth of a river where fresh and saline water get mixed) and the Arctic, the salinity fluctuates from 0 – 35, seasonally (fresh water coming from ice caps).

Pacific Ocean

- The salinity variation in the Pacific Ocean is mainly due to its shape and larger areal extent.

Atlantic Ocean

- The average salinity of the Atlantic Ocean is around 36-37.

- The equatorial region of the Atlantic Ocean has a salinity of about 35.

- Near the equator, there is heavy rainfall, high relative humidity, cloudiness and calm air of the doldrums.

- The polar areas experience very little evaporation and receive large amounts of fresh water from the melting of ice. This leads to low levels of salinity, ranging between 20 and 32.

- Maximum salinity (37) is observed between 20 °N and 30 °N and 20 °W – 60 °W. It gradually decreases towards the north.

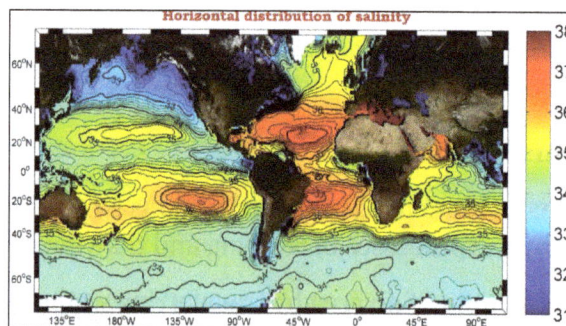

Horizontal distribution of salinity

Indian Ocean

- The average salinity of the Indian Ocean is 35.

- The low salinity trend is observed in the Bay of Bengal due to influx of river water by the river Ganga.

- On the contrary, the Arabian Sea shows higher salinity due to high evaporation and low influx of fresh water.

Marginal Seas

- The North Sea, in spite of its location in higher latitudes, records higher salinity due to more saline water brought by the North Atlantic Drift.

- Baltic Sea records low salinity due to influx of river waters in large quantity.

- The Mediterranean Sea records higher salinity due to high evaporation.

- Salinity is, however, very low in Black Sea due to enormous fresh water influx by rivers.

Inland Seas and Lakes

- The salinity of the inland Seas and lakes is very high because of the regular supply of salt by ' the rivers falling into them.

- Their water becomes progressively more saline due to evaporation.

- For instance, the salinity of the Great Salt Lake, (Utah, USA), the Dead Sea and the Lake Van in Turkey is 220, 240 and 330 respectively.

- The oceans and salt lakes are becoming more salty as time goes on because the rivers dump more salt into them, while fresh water is lost due to evaporation.

Cold and Warm Water Mixing Zones

Salinity decreases from 35 – 31 on the western parts of the northern hemisphere because of the influx of melted water from the Arctic region.

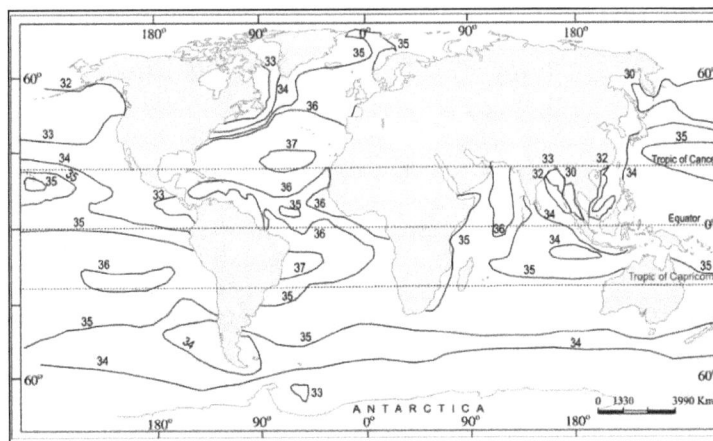

Surface salinity of the world's oceans.

Sub-Surface Salinity

- With depth, the salinity also varies, but this variation again is subject to latitudinal difference. The decrease is also influenced by cold and warm currents.

- In high latitudes, salinity increases with depth. In the middle latitudes, it increases up to 35 metres and then it decreases. At the equator, surface salinity is lower.

Vertical Distribution of Salinity

- Salinity changes with depth, but the way it changes depends upon the location of the sea.

- Salinity at the surface increases by the loss of water to ice or evaporation, or decreased by the input of fresh waters, such as from the rivers.

- Salinity at depth is very much fixed, because there is no way that water is 'lost', or the salt is 'added.' There is a marked difference in the salinity between the surface zones and the deep zones of the oceans.

- The lower salinity water rests above the higher salinity dense water.

- Salinity, generally, increases with depth and there is a distinct zone called the halocline (compare this with thermocline), where salinity increases sharply.

- Other factors being constant, increasing salinity of seawater causes its density to increase. High salinity seawater, generally, sinks below the lower salinity water. This leads to stratification by salinity.

Movements in the Ocean

There are different types of movements of ocean water under the influence of different physical characteristics like temperature, salinity, density, etc. Movements of ocean water are also affected by external forces like the sun, moon and the winds.

The major movements of the ocean waters can be classified into three. They are:

1. Waves

2. Tides

3. Ocean Currents

Waves and the ocean currents are horizontal movements of ocean waters while the tide is a kind of vertical movement of the ocean water.

Waves

- Waves are nothing but the oscillatory movements that result in the rise and fall of water surface.

- Waves are a kind of horizontal movements of the ocean water.

- They are actually the energy, not the water as such, which moves across the ocean surface.

- This energy for the waves is provided by the wind.

- In a wave, the movement of each water particle is in a circular manner.

- A wave has two major parts: the raised part is called as the crest while the low-point is called as the trough.

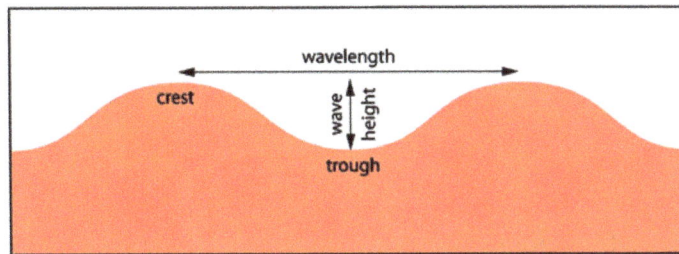

Tides

- Tide are the periodical rise and fall of the sea levels, once or twice a day, caused by the combined effects of the gravitational forces exerted by the sun, the moon and the rotation of the earth.

- They are a vertical movement of waters and are different from movements of ocean water caused by meteorological effects like the winds and atmospheric pressure changes.

- Note: The water movements which are caused by the meteorological effects like the said above are called as surges and they are not regular like tides.

- The moon's gravitational pull to a great extent is the major cause of the occurrence of tides (the moon's gravitational attraction is more effective on the earth than that of the sun).

- Sun's gravitational pull and the centrifugal force due to the rotation of earth are the other forces which act along with the moon's gravitational pull.

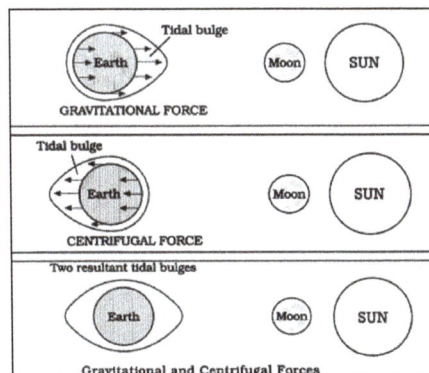

- The highest tides in the world occur in the Bay of Fundi in Canada.

- When the tide is channeled between islands or into bays and estuaries, they are termed as Tidal Currents.

- The regular interval between two high or two low tides is 12 hours 25 minutes.

Flow Tide and Ebb Tide

- A flow tide or a flood tide is a rising tide or incoming tide which results in a high tide.

- It is thus the time period between a low tide and a high tide (i.e., the rising time).

- Ebb Tide is the receding or outgoing tide. It is the period between high tide and low tide during which water flows away from the shore.

Types of Tides

A. Tides Based on the Frequency

- Semi-diurnal Tide: They are the most common tidal pattern, featuring two high tides and two low tides each day.

- Diurnal Tides: Only one high tide and one low tide each day.

- Mixed Tide: Tides having variations in heights are known as mixed tides. They generally occur along the west coast of North America.

B. Tides Based on the Sun, the Moon, and the Earth's Positions

- Spring Tides: When the sun, the moon, and the earth are in a straight line, the height of the tide will be higher than normal. These are called as a spring tides. They occur twice in a month-one on the full moon and the other on the new moon.

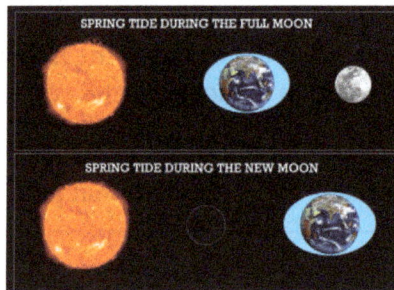

- Neap Tides: Normally after seven days of a spring tide, the sun and the moon become at a right angle to each other with respect to the earth. Thus, the gravitational forces of the sun and the moon tend to counteract one another. The tides during this period will be lower than the normal which are called as the neap tides. They also occur twice in a month-during the first quarter moon and the last quarter moon.

Magnitude of Tides

- Perigee: When the moon's orbit is closest to the earth, it is called as perigee. During this period, unusually high and low tide occur.

- Apogee: When the moon's orbit is farthest from the earth, it is called as apogee. Tidal ranges will be much less than the average during this period.

- Perihelion: It is the position where the earth is closest to the sun (around January 3rd). Unusual high and low tides occur during this time.

- Aphelion: It is the position where the earth is farthest from the sun (around July 4th). Tidal ranges are much less than the average during this period.

Tidal Bore

Tidal bore.

When the leading edge of the incoming tide forms a wave/waves of water that travel up a river or a narrow bay against the direction of the river or bay's current, it is called as a tidal bore. The Indian rivers like the Ganges, Brahmaputra, Indus, etc. exhibits tidal bores.

Inter-Tidal Zone

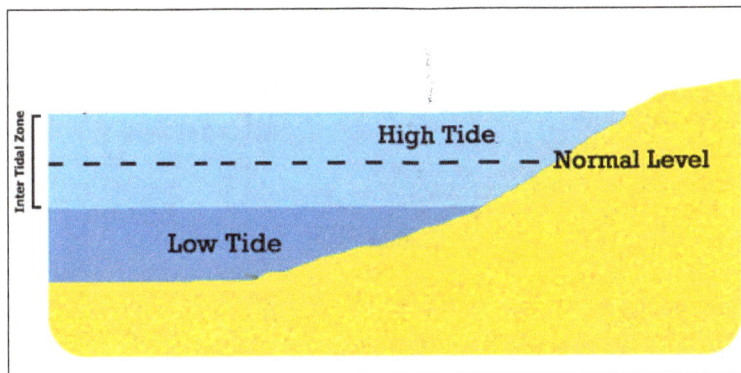

The intertidal zone, also known as the foreshore and seashore and sometimes referred to as the littoral zone, is the area that is above water at low tide and under water at high tide (i.e., the area between the tide-marks).

Effects of Tides

- Tides act as a link between the port and the open sea. Some of the major ports of the world, such as London port on the river Thames and Kolkata port on river Hugli are located on the rivers away from the sea coast.

- The tidal current clear away the river sediments and slows down the growth of delta.

- It increases the depth of water which helps ships to move safely to the ports.

- It also acts as a source for producing electricity.

Ocean Currents

- The ocean currents are the horizontal flow of a mass of water in a fairly defined direction over great distances.

- They are just like a river flowing in an ocean.

- Ocean currents can be formed by the winds, density differences in ocean waters due to differences in temperature and salinity, gravity and events such as earthquakes.

- The direction of movement of an ocean current is mainly influenced by the rotation of the earth (due to Coriolis force, most ocean currents in northern hemisphere move in clockwise manner and ocean currents in southern hemisphere move in an anti-clockwise manner).

Gyre, Drift and Stream

- Any large system of rotating ocean current, particularly those involved with large wind movements is called as a Gyre. They are caused by the Coriolis force.

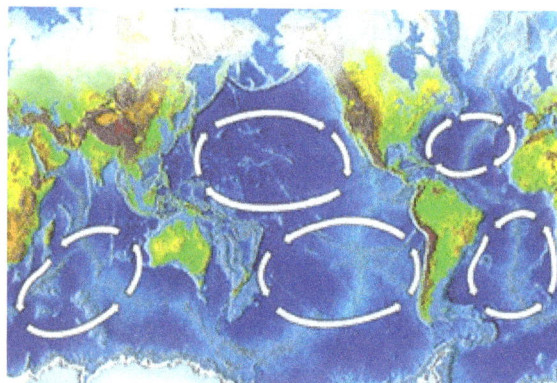

- When the ocean water moves forward under the influence of prevailing wind, it is called as Drift (The term 'drift' is also used to refer the speed of an ocean current which is measured in knots). E.g. North Atlantic Drift.

- When a large mass of the ocean water moves in a definite path just like a large river on the continent, it is called as a Stream. They will have greater speed than drifts. E.g. Gulf Stream.

Types of Ocean Currents

Warm Ocean Currents

- Those currents which flow from equatorial regions towards poles which have a higher surface temperature and are called warm current.

- They bring warm waters to the cold regions.

- They are usually observed on the east coast of the continents in the lower and middle latitudes of both hemispheres.

- In the northern hemisphere, they are also found on the west coast of the continents in the higher latitudes (E.g. Alaska and Norwegian Currents).

Cold Ocean Currents

- Those currents which flow from polar regions towards equator have a lower surface temperature and are called cold currents.

- They bring cold waters into warm areas.

- These currents are usually found on the west coast of the continents in low and middle latitudes of both hemispheres.

- In the northern hemisphere, they are also found on the east coast in the higher latitudes (E.g. Labrador, East Greenland and Oyashio currents).

The ocean currents can be also classified as:

- Surface Currents: They constitute about 10% of all the waters in an ocean. These waters are occupied at the upper 400m of an ocean or the Ekman Layer. It is the layer of the ocean water which moves due to the stress of blowing the wind and this motion is thus called as Ekman Transport.

- Deep Water Currents: They constitute about 90% of the ocean water. They move around the ocean basin due to variations in the density and gravity.

Factors influencing the Origin and Nature of Ocean Currents

Difference in Density

- As we all know, the density of sea water varies from place to place according to its temperature and proportion of salinity.

- The density increases with an increase in salinity and decreases with a decrease in salinity.

- But when the temperature increases, density decreases and when the temperature decreases density increases.

- This increase and decrease in density due to the differences in temperature and salinity causes the water to move from one place to another.

- Such a movements of water due to the differences in density as a function of water temperature and salinity is called as the Thermohaline Circulation.

- In polar regions, due to a lower temperature, the waters will be of high density. This causes the waters to sink to the bottom and then to move towards the less dense middle and lower latitudes (or towards the equatorial regions).

- They rise (upwelling) at the warm region and push the already existing less dense, warm water towards the poles.

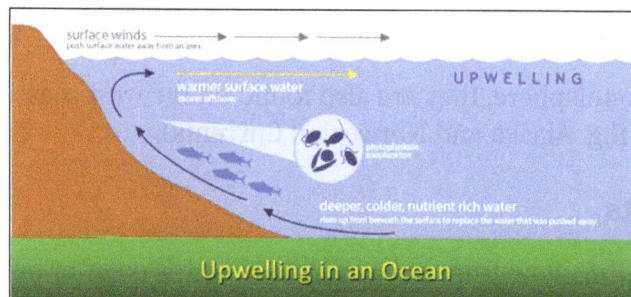

Upwelling in an Ocean

While considering the equatorial region, the high temperature in those regions causes the water to expand. Thus, the waters in these regions will be at a higher level than that of the middle and upper latitudes. This also creates a gradient and results in the movement of waters from equatorial region to middle and upper latitudes.

The Earth's Rotation

- Earth's rotation causes Coriolis force which deflects the air to its right in the northern hemisphere and to its left in the southern hemisphere-Ferrel's Law.

- Similarly, oceans water also affected by the Coriolis force and follows the Ferrel's Law.

- Hence, ocean currents in the northern hemisphere move in a clockwise (towards right) direction and ocean currents in southern hemisphere moves in an anti-clockwise (towards left) direction (In the Indian Ocean due to the impact of the Asian monsoon, the currents in the northern hemisphere do not follow this pattern of movements all time).

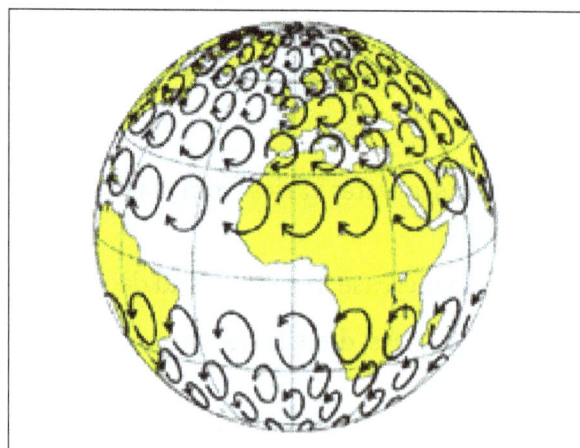

The Winds

- The winds like trade winds and westerlies drive the ocean water in a steady flow in front of them.

- When the direction of the winds changes, the direction of the current also gets changed.

References

- Different-steps-of-the-hydrologic-cycle: conserve-energy-future.com, Retrieved 2 August, 2019

- Temperature-of-oceanic-water-oceans-geography, oceanography: geographynotes.com, Retrieved 12 June, 2019

- Temperature-distribution-of-oceans, factors-affecting-vertical-temperature-distribution, horizontal-temperature-distribution: pmfias.com, Retrieved 21 February, 2019

- Ocean-salinity-vertical-horizontal-distribution: pmfias.com, Retrieved 1 July, 2019

- Movements-ocean-waves-tides-currents: clearias.com, Retrieved 31 January, 2019

Permissions

We would like to thank the editorial team for lending their expertise to make the book truly unique. They have played a crucial role in the development of this book. Without their invaluable contributions this book wouldn't have been possible. They have made vital efforts to compile up to date information on the varied aspects of this subject to make this book a valuable addition to the collection of many professionals and students.

This book was conceptualized with the vision of imparting up-to-date and integrated information in this field. To ensure the same, a matchless editorial board was set up. Every individual on the board went through rigorous rounds of assessment to prove their worth. After which they invested a large part of their time researching and compiling the most relevant data for our readers.

The editorial board has been involved in producing this book since its inception. They have spent rigorous hours researching and exploring the diverse topics which have resulted in the successful publishing of this book. They have passed on their knowledge of decades through this book. To expedite this challenging task, the publisher supported the team at every step. A small team of assistant editors was also appointed to further simplify the editing procedure and attain best results for the readers.

Apart from the editorial board, the designing team has also invested a significant amount of their time in understanding the subject and creating the most relevant covers. They scrutinized every image to scout for the most suitable representation of the subject and create an appropriate cover for the book.

The publishing team has been an ardent support to the editorial, designing and production team. Their endless efforts to recruit the best for this project, has resulted in the accomplishment of this book. They are a veteran in the field of academics and their pool of knowledge is as vast as their experience in printing. Their expertise and guidance has proved useful at every step. Their uncompromising quality standards have made this book an exceptional effort. Their encouragement from time to time has been an inspiration for everyone.

The publisher and the editorial board hope that this book will prove to be a valuable piece of knowledge for students, practitioners and scholars across the globe.

Index